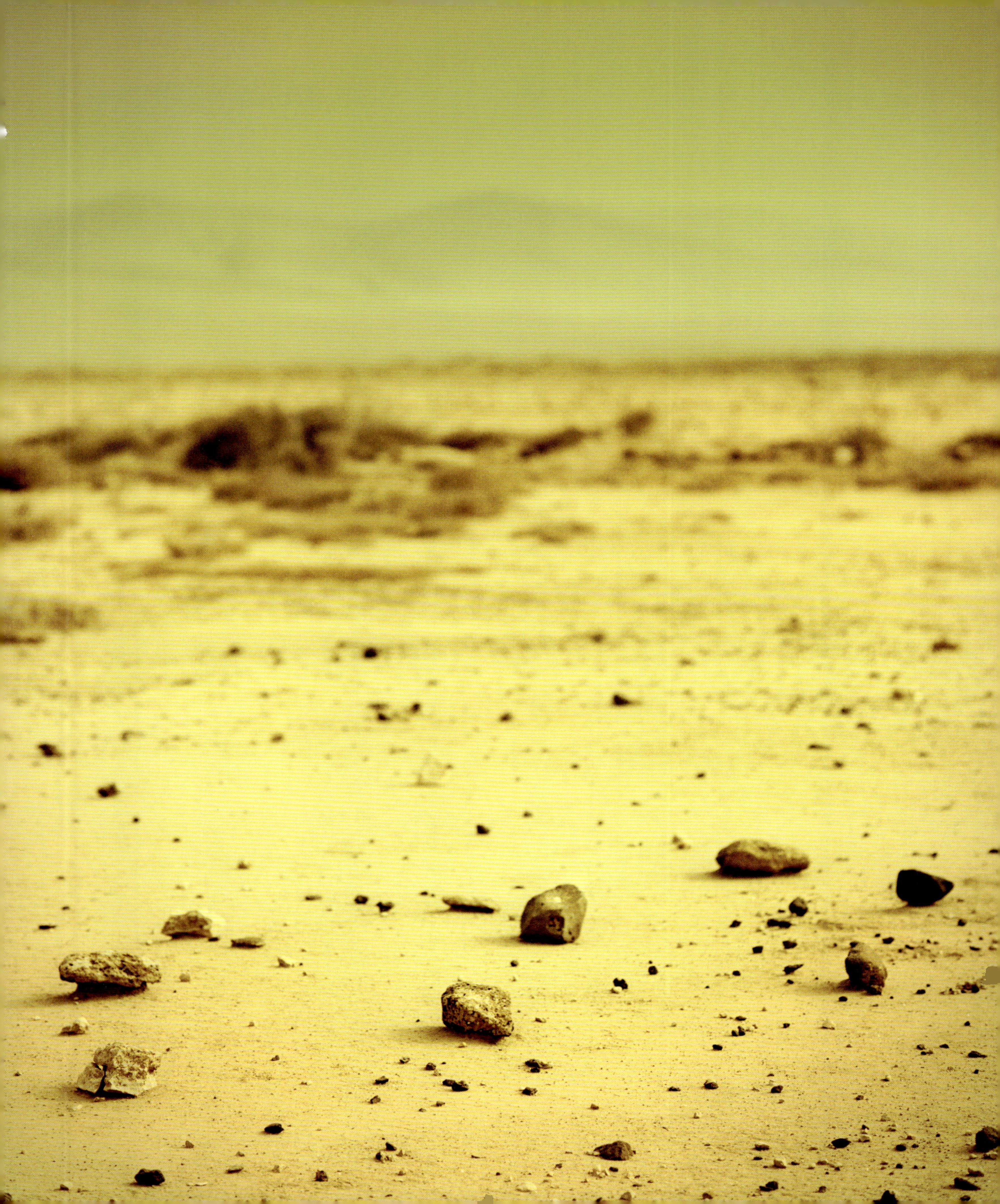

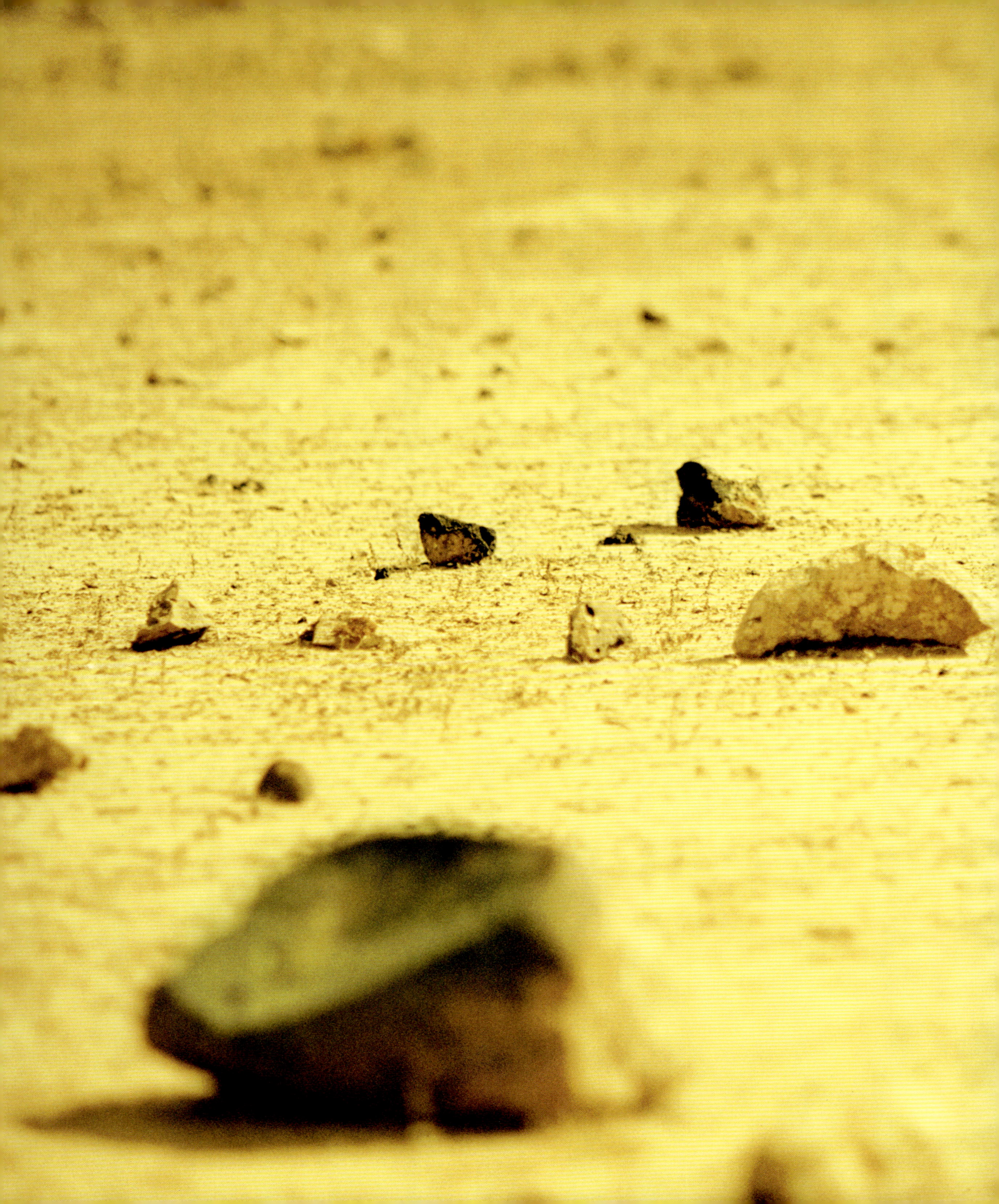

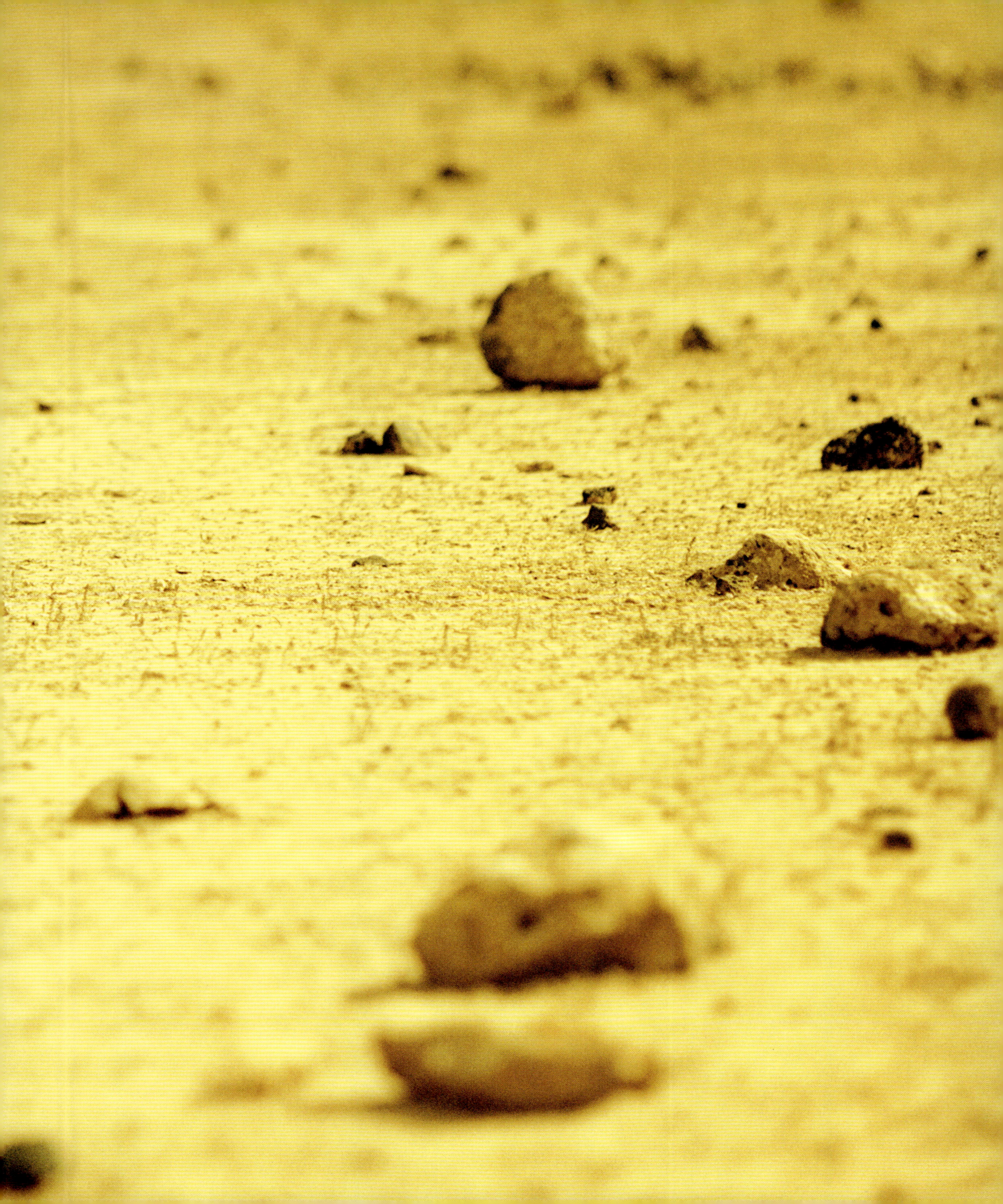

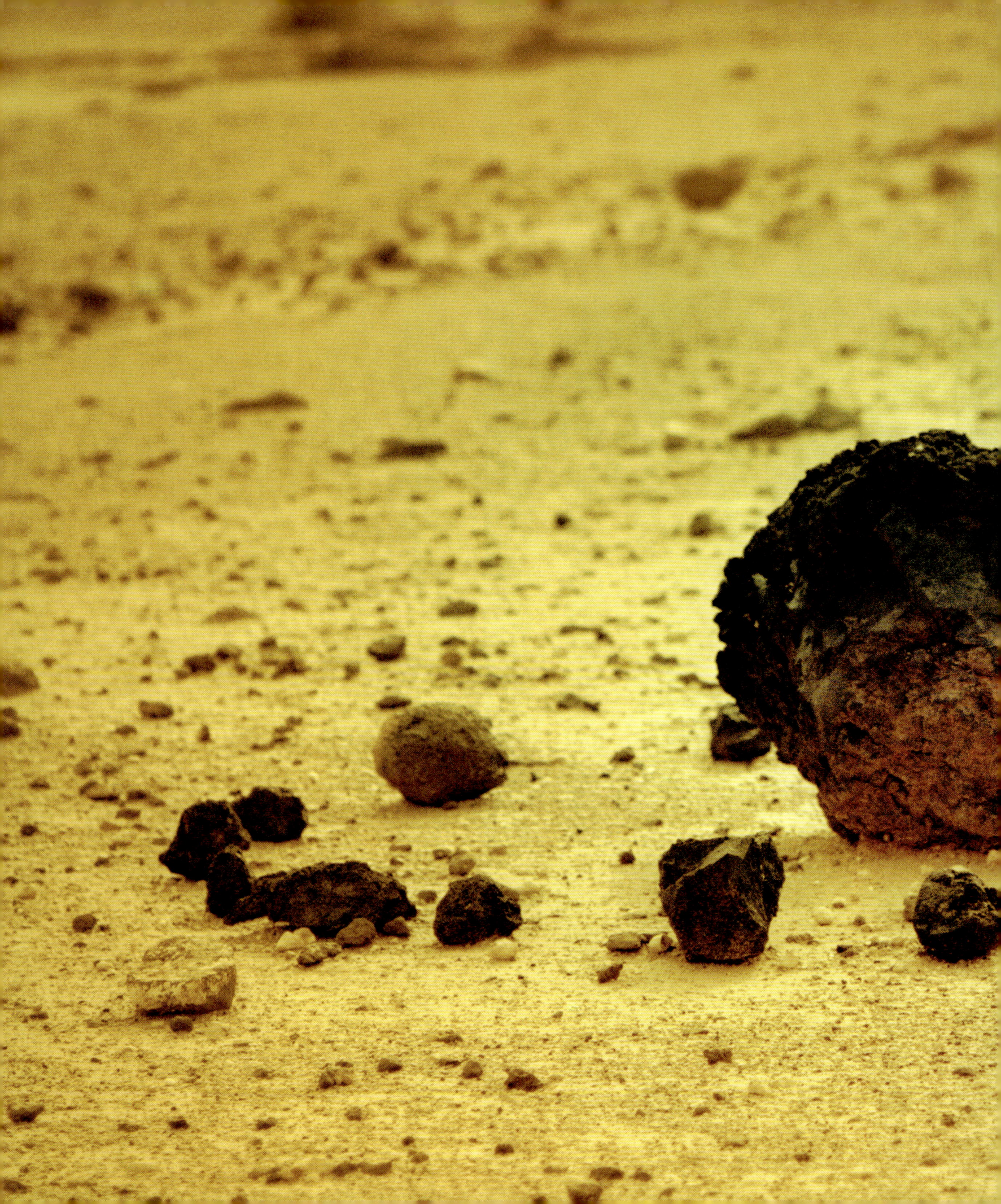

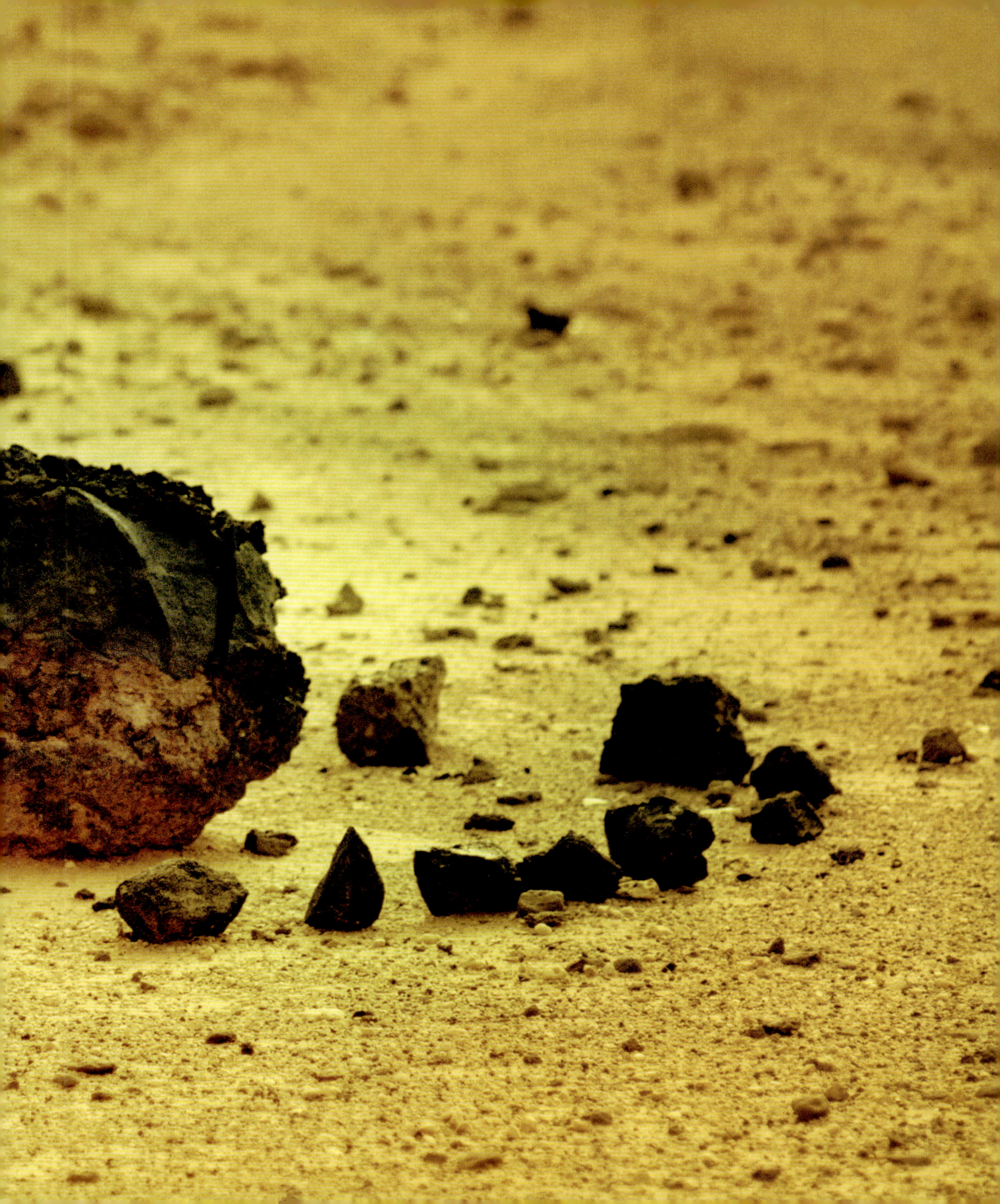

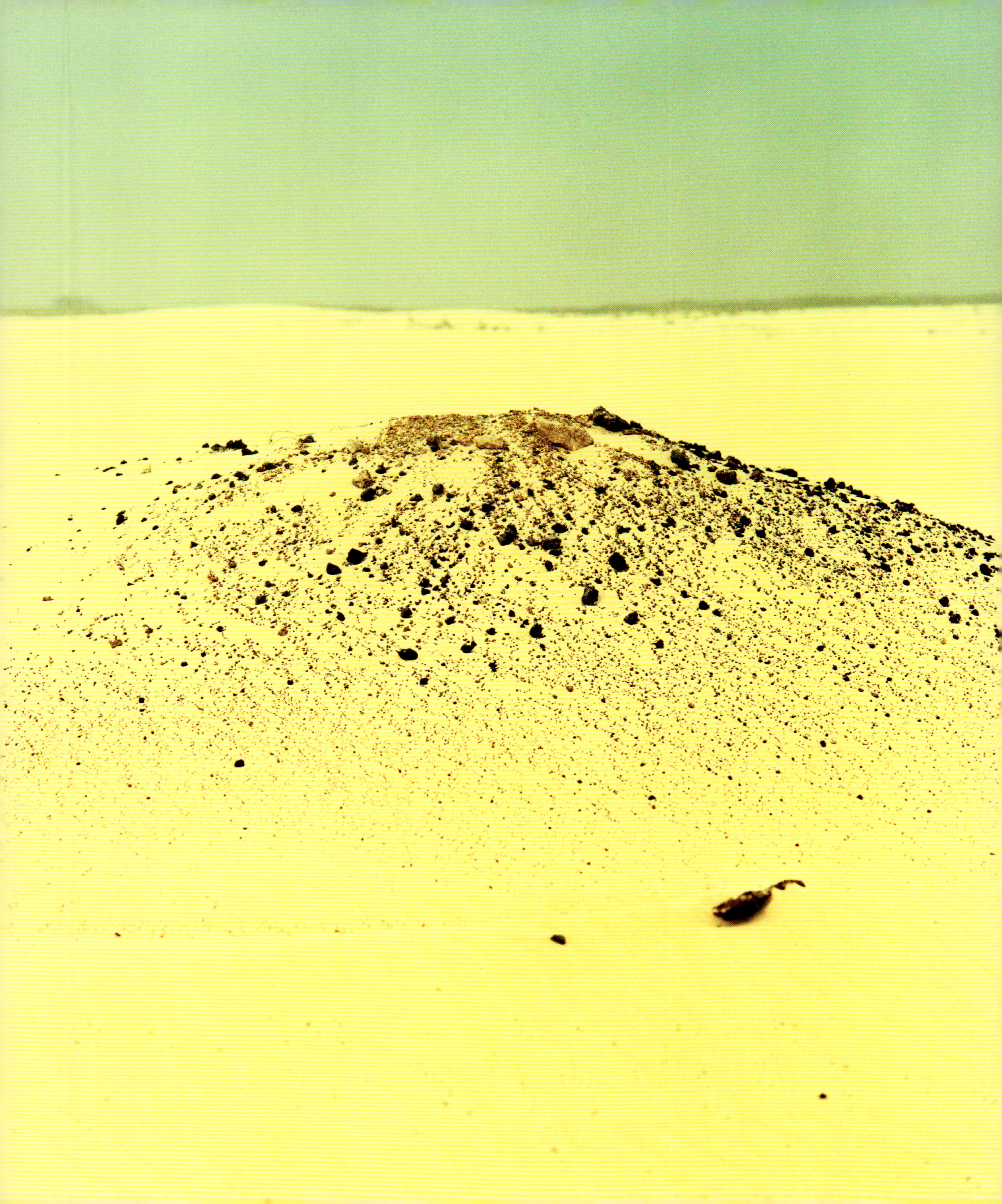

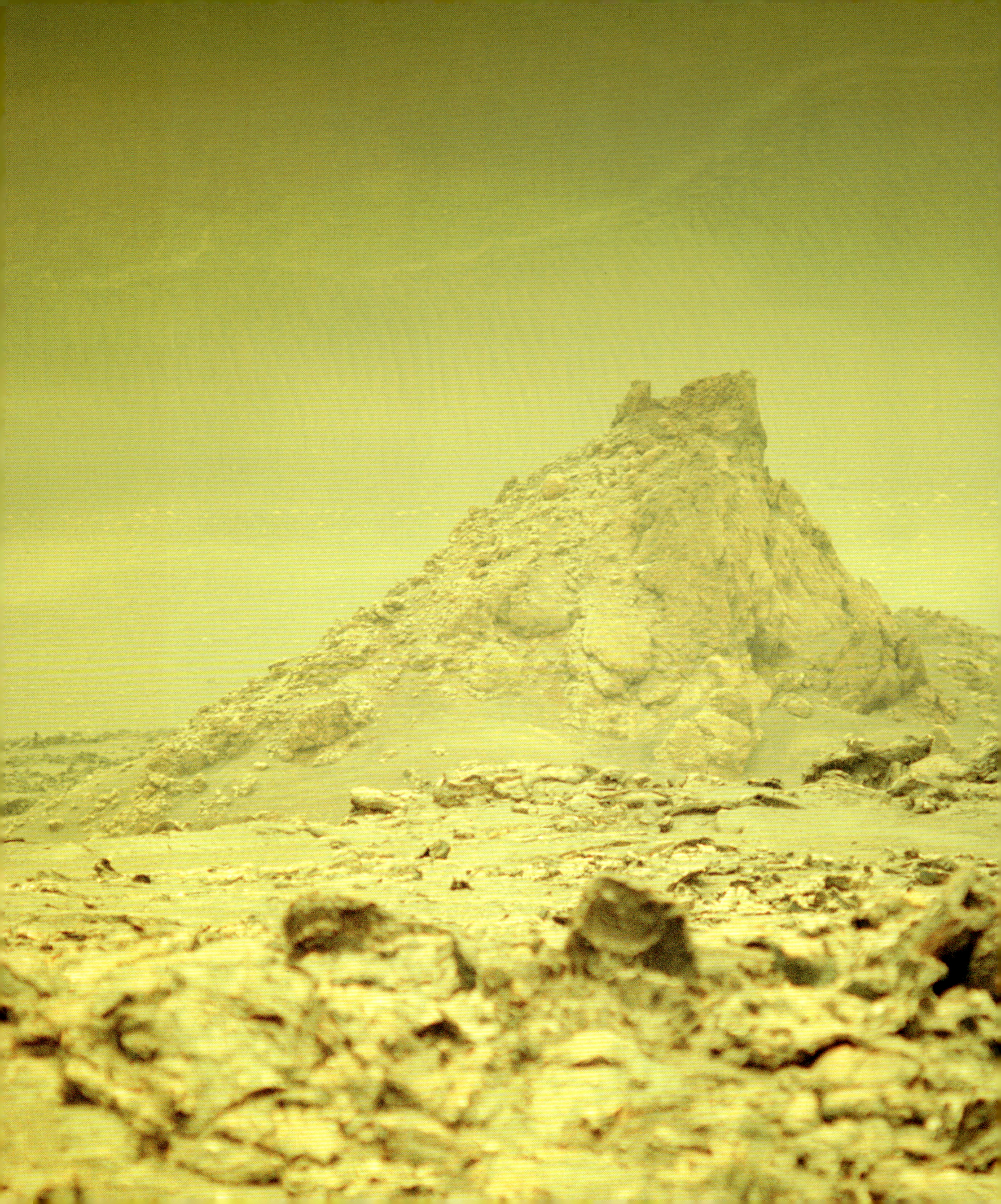

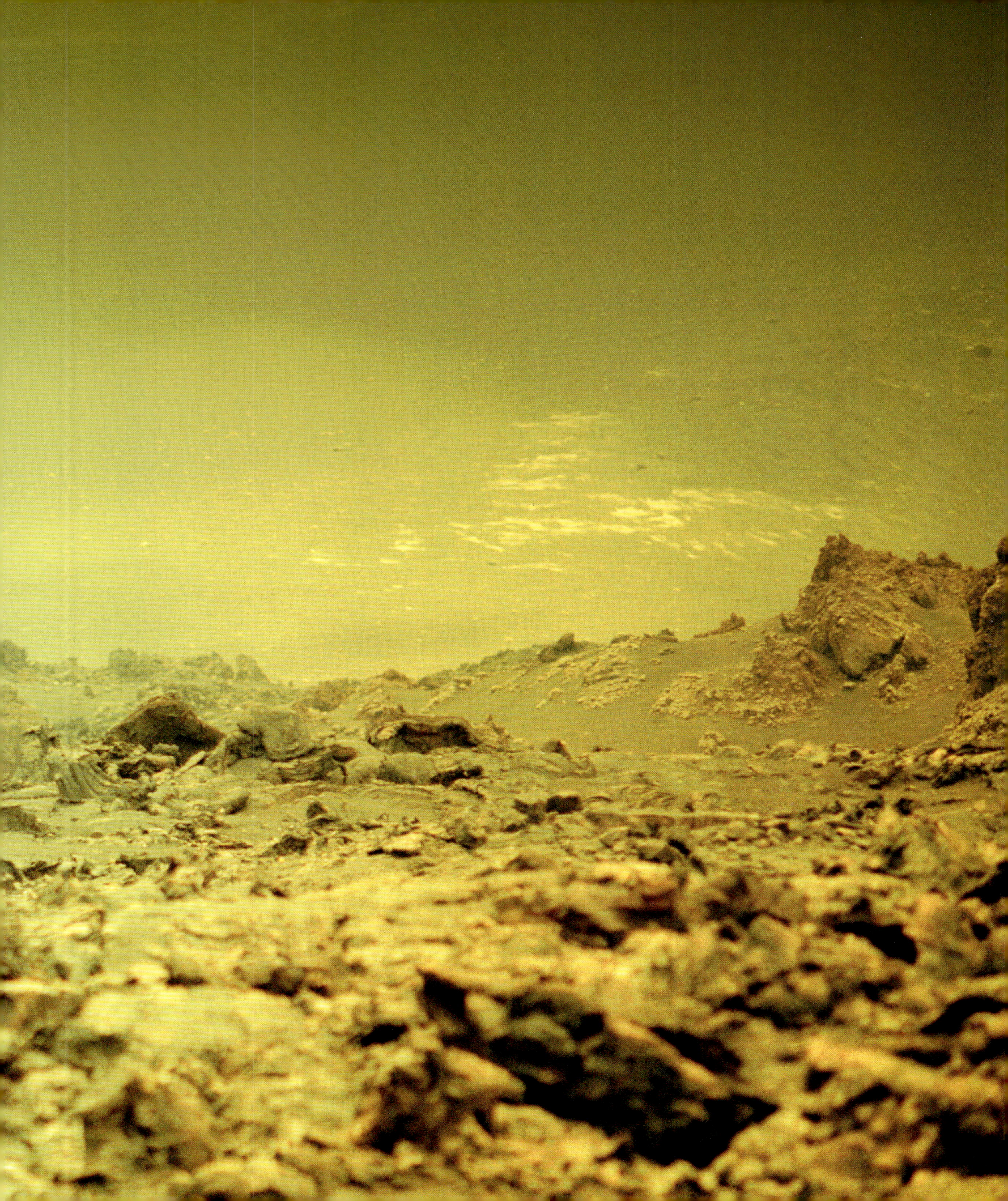

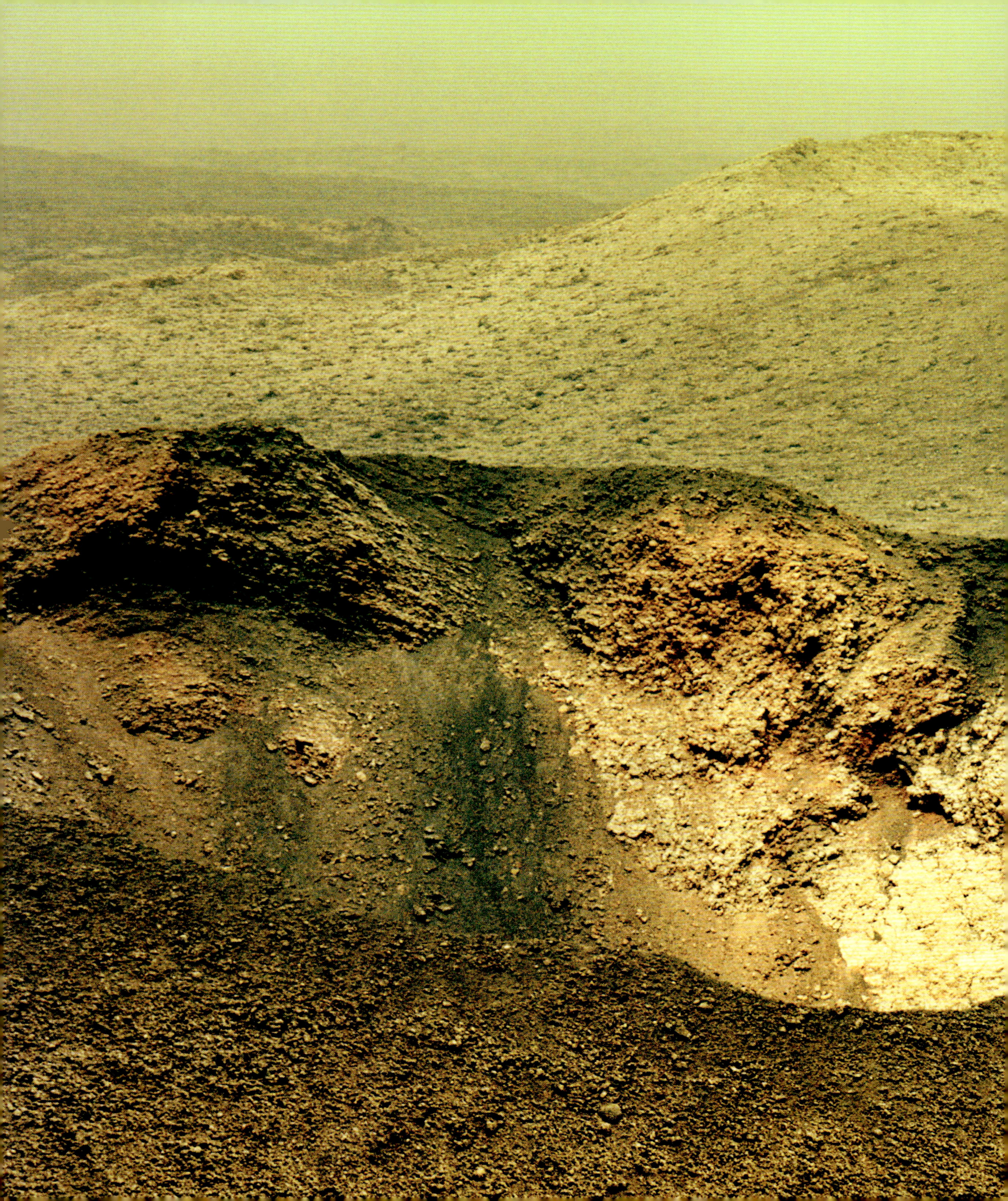

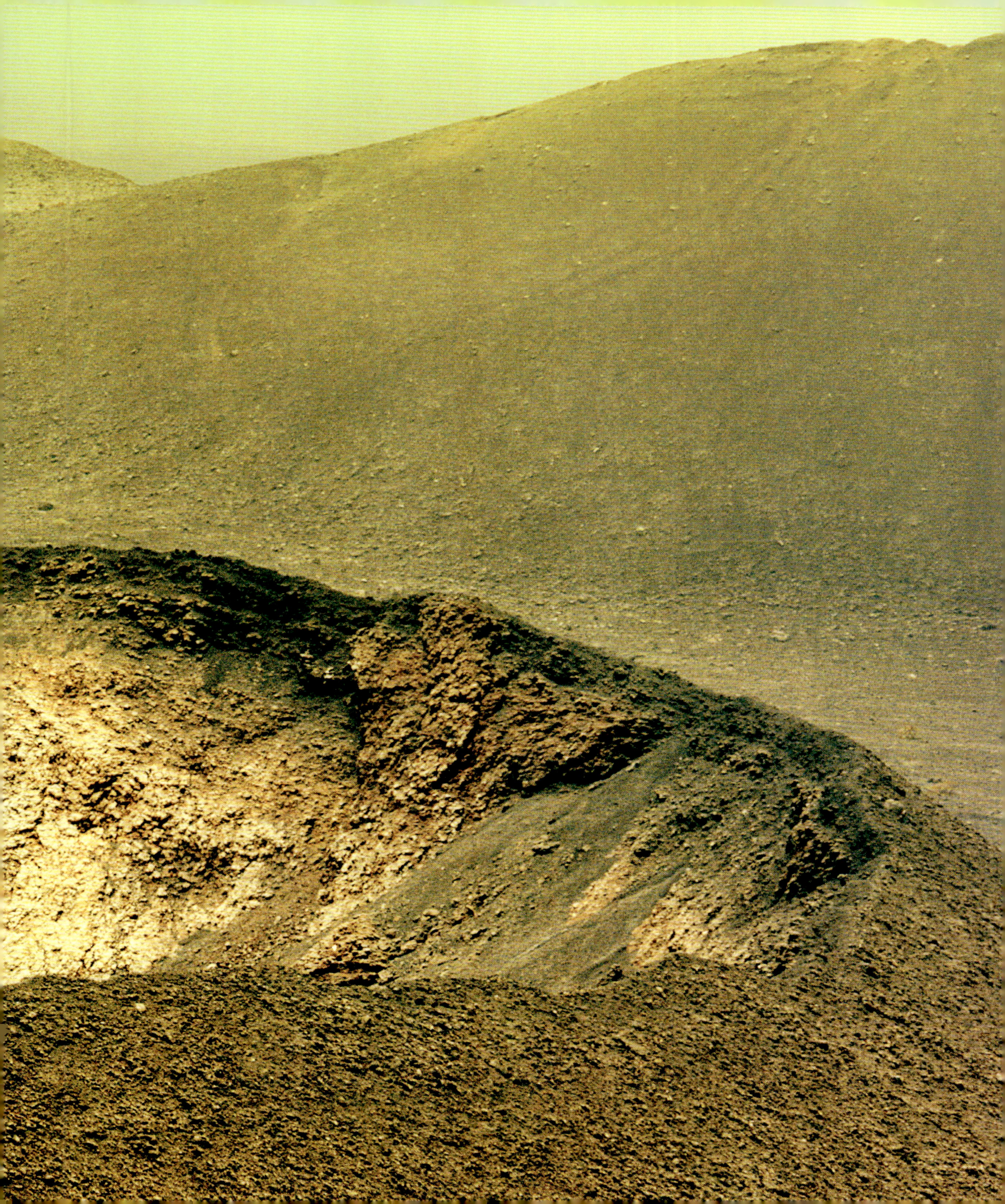

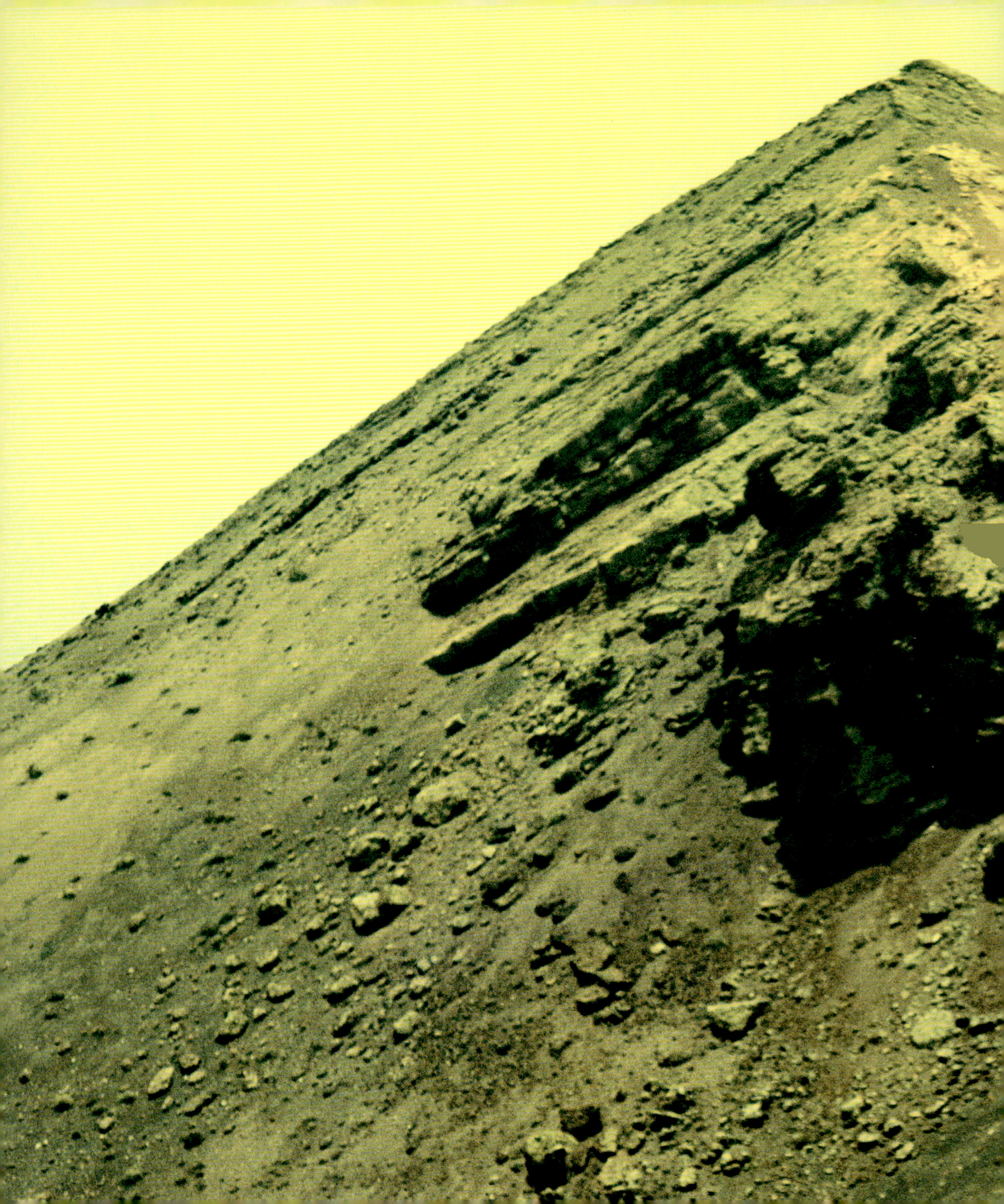

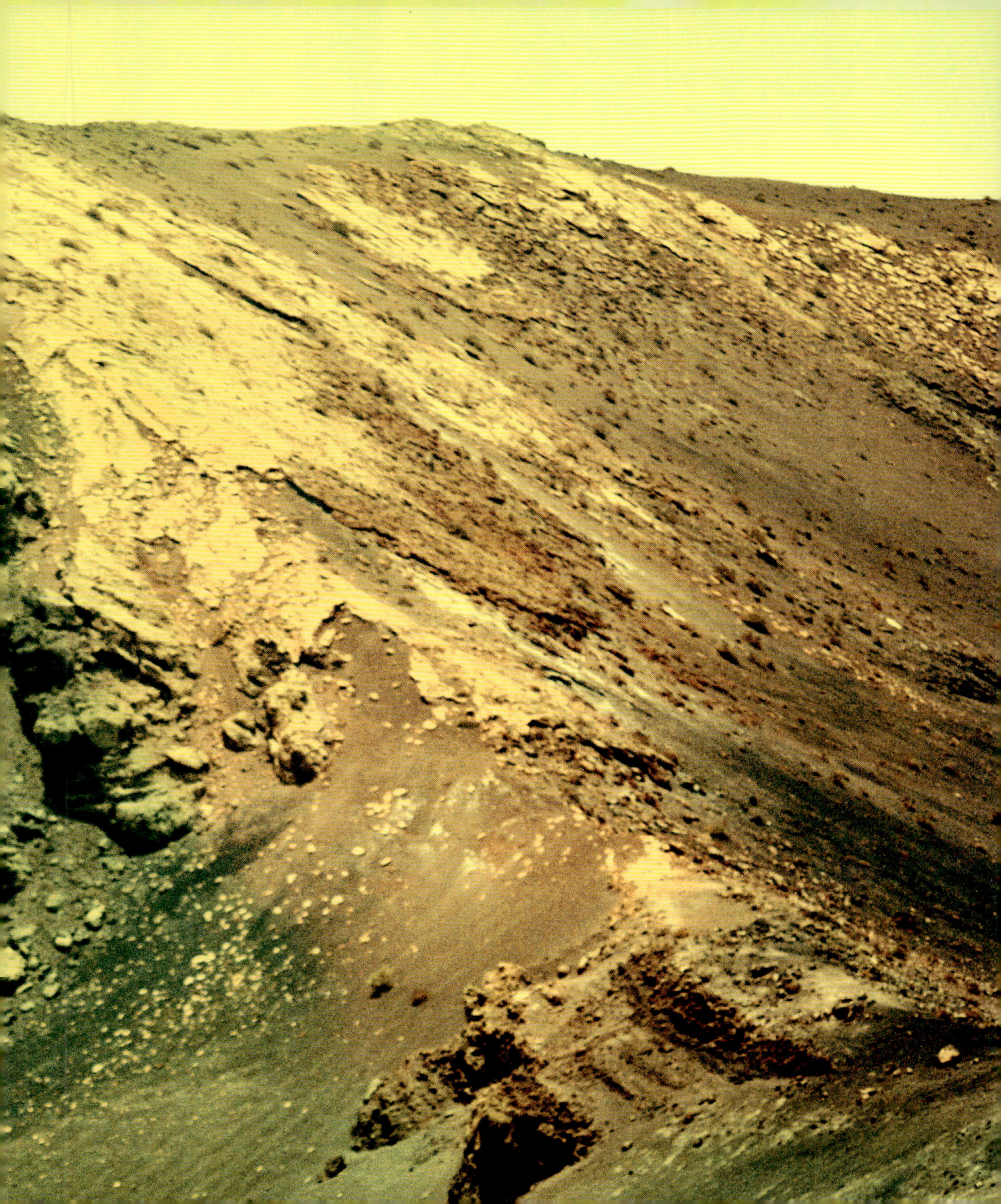

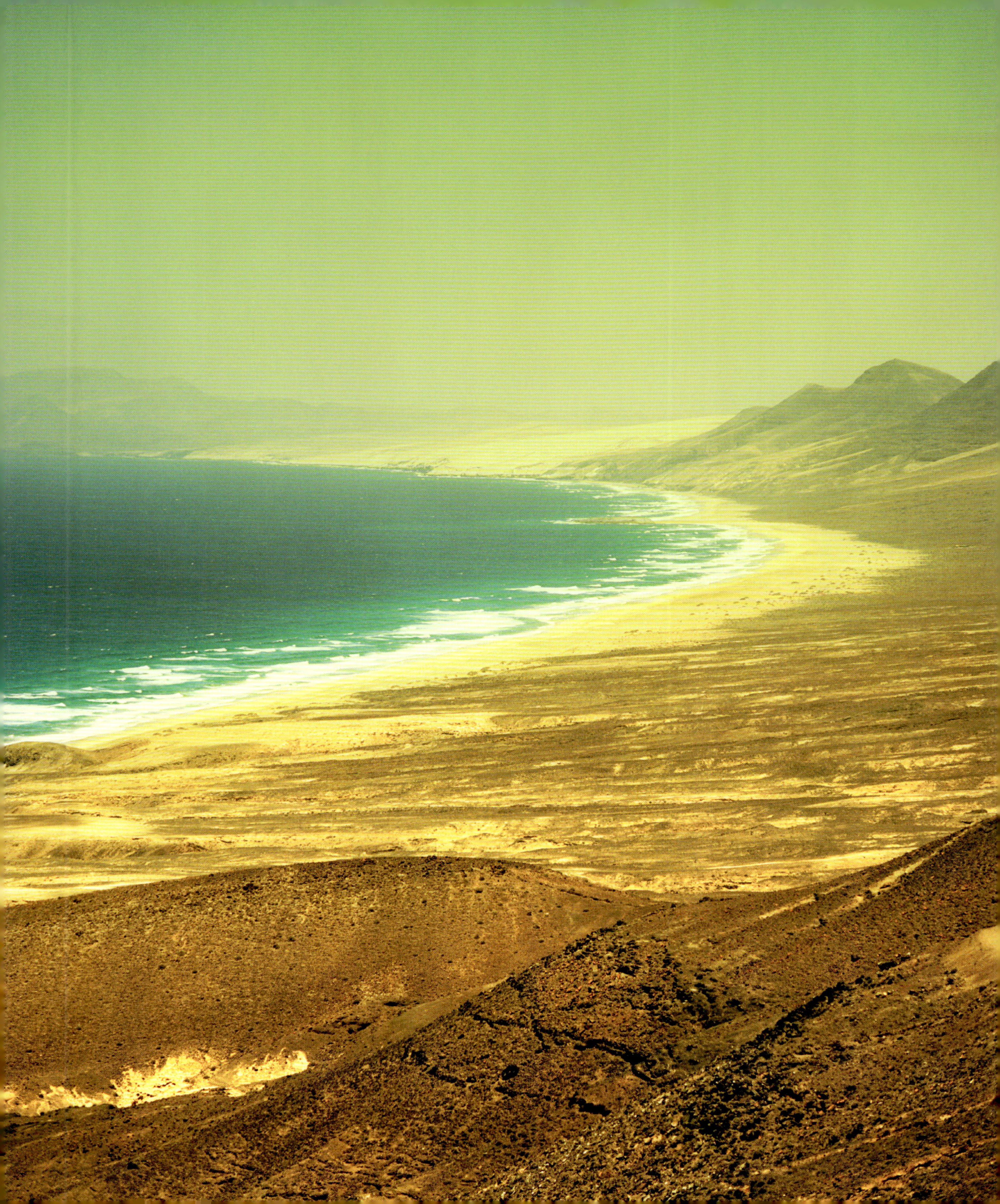

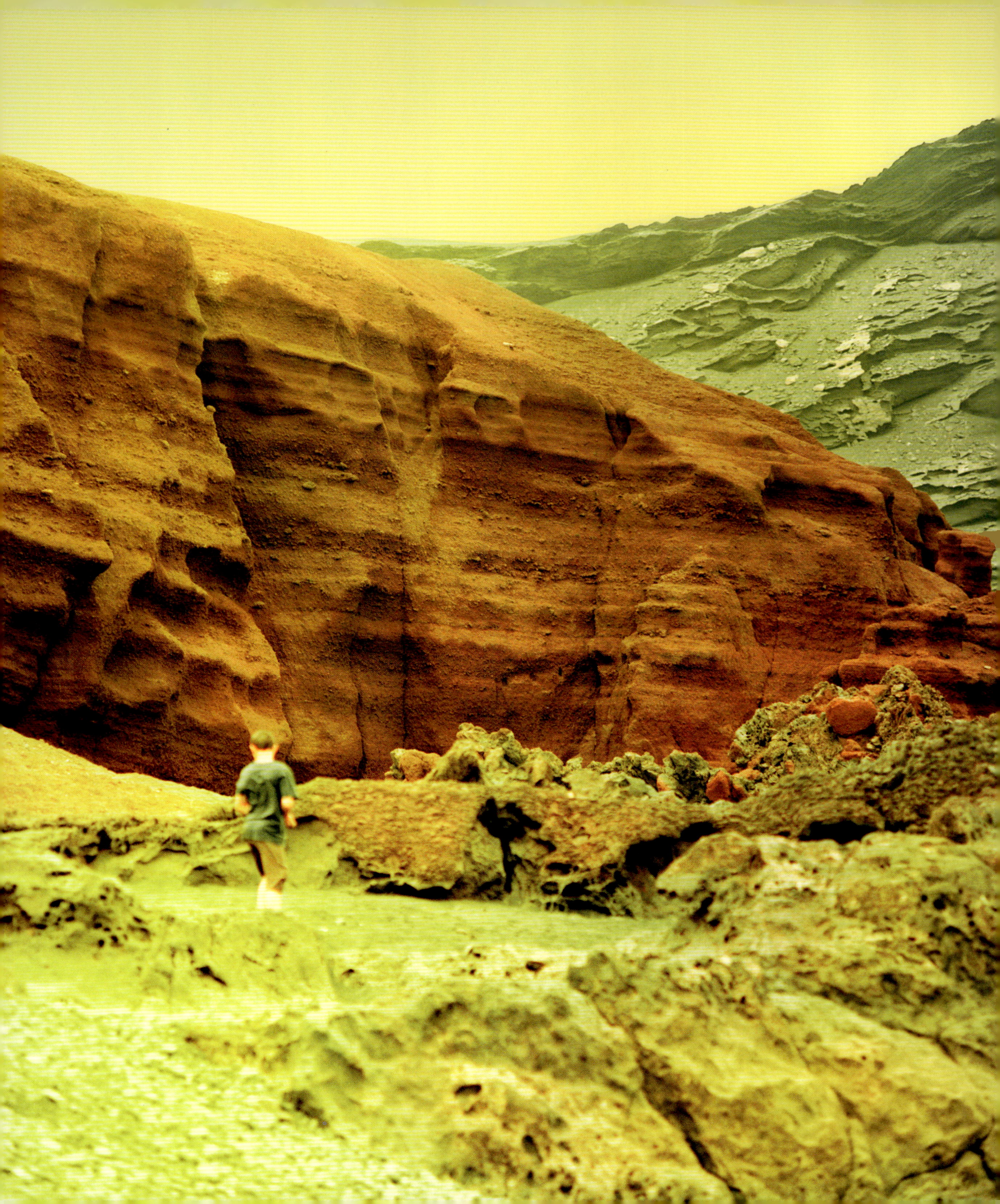

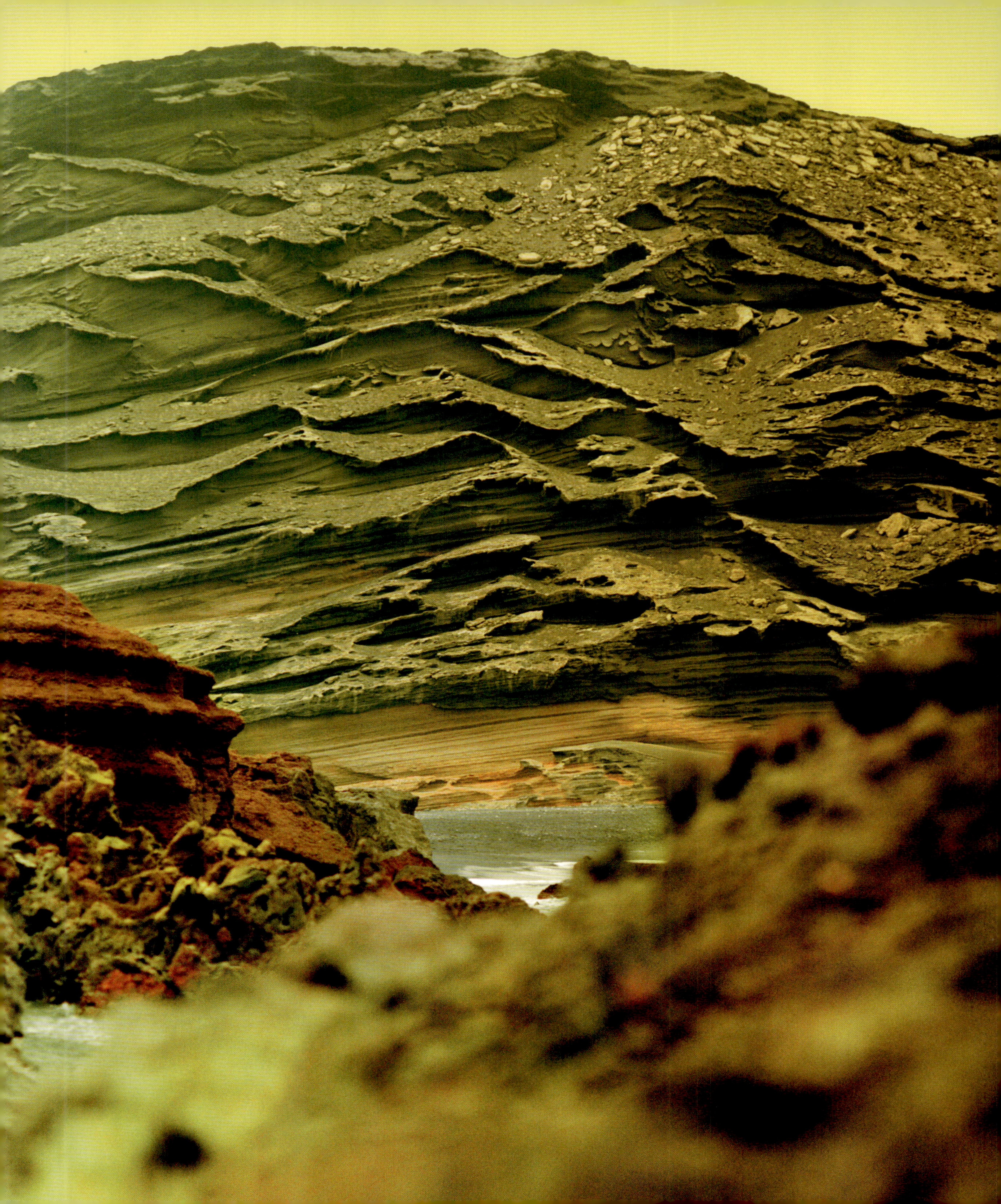

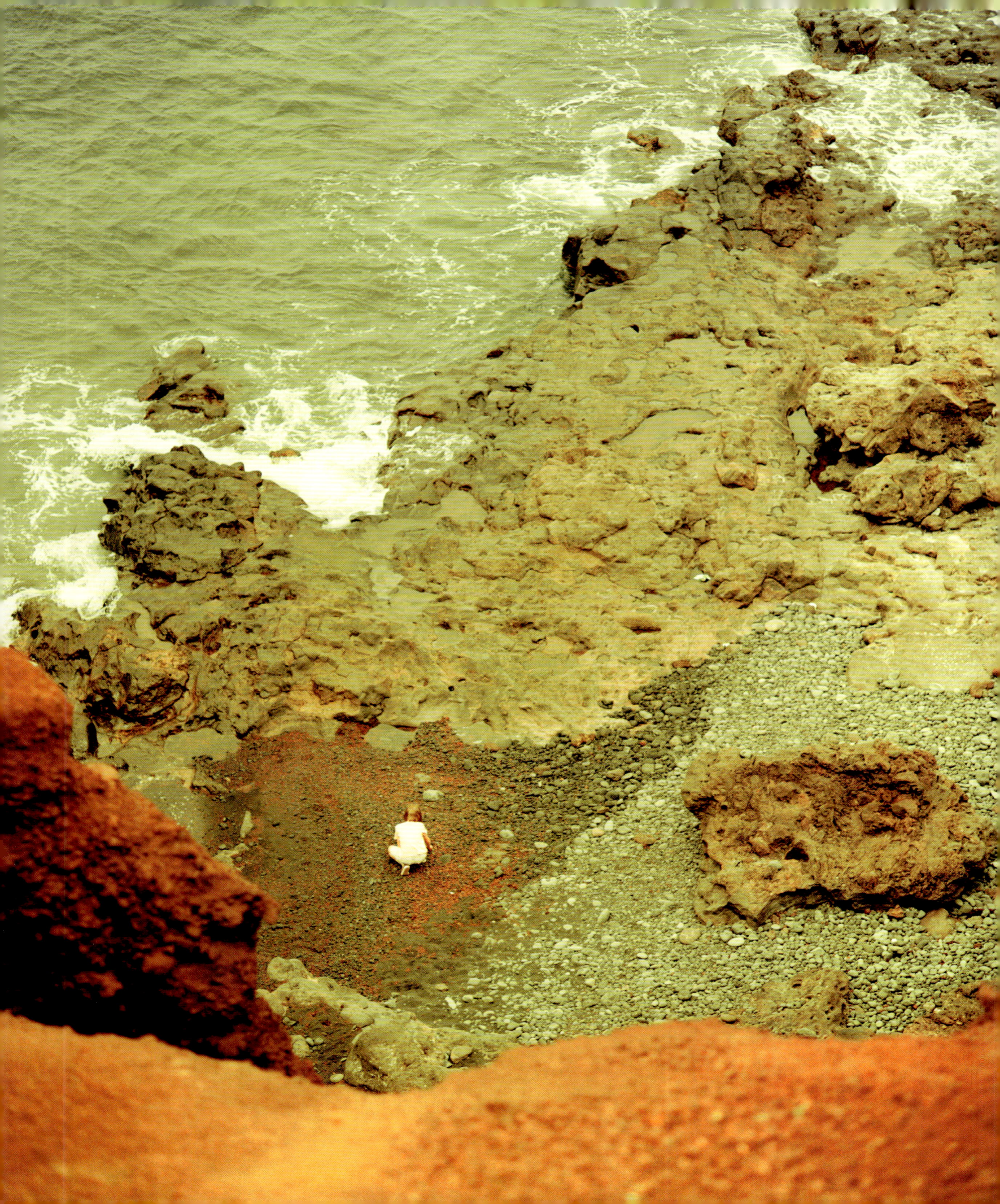

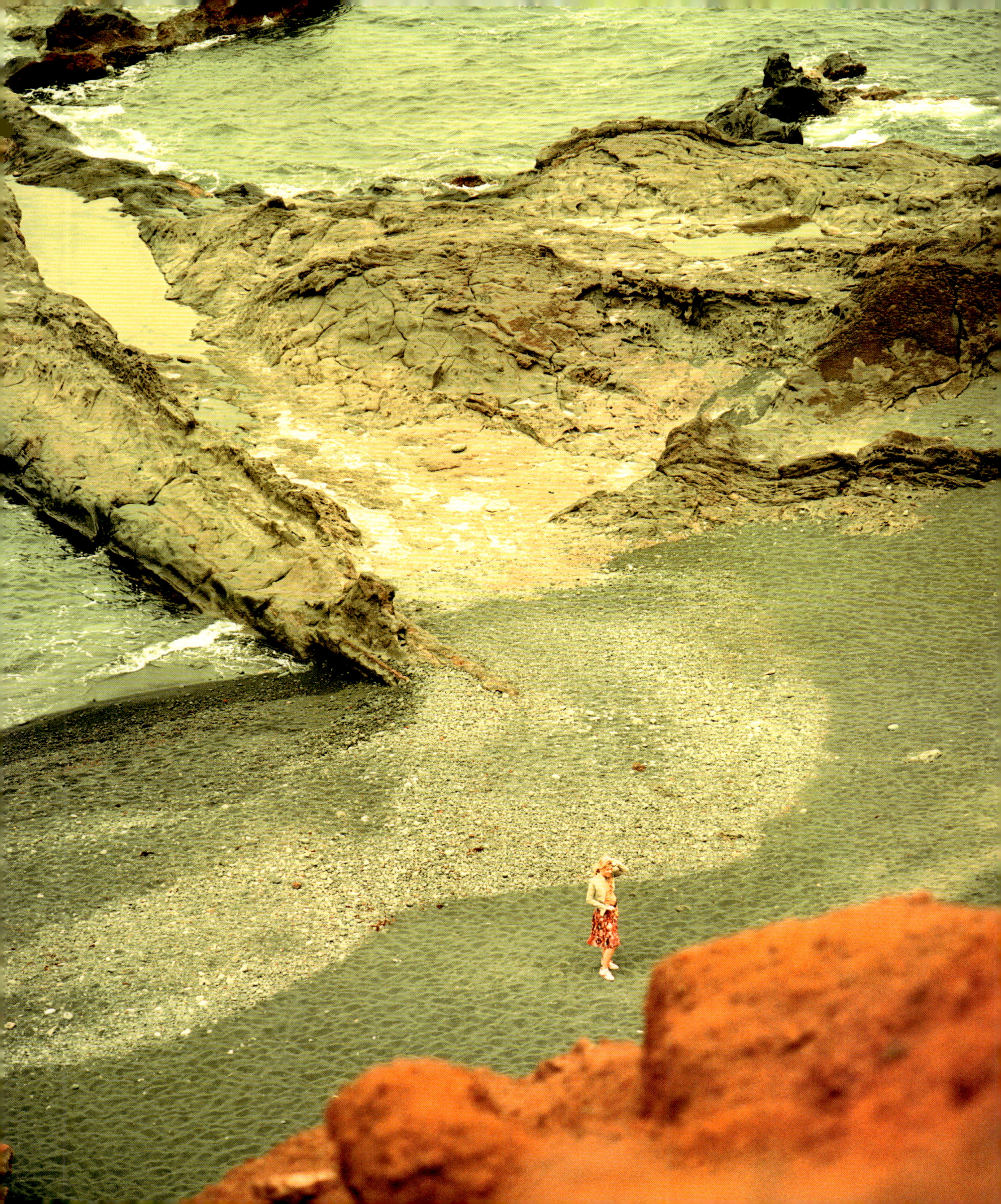

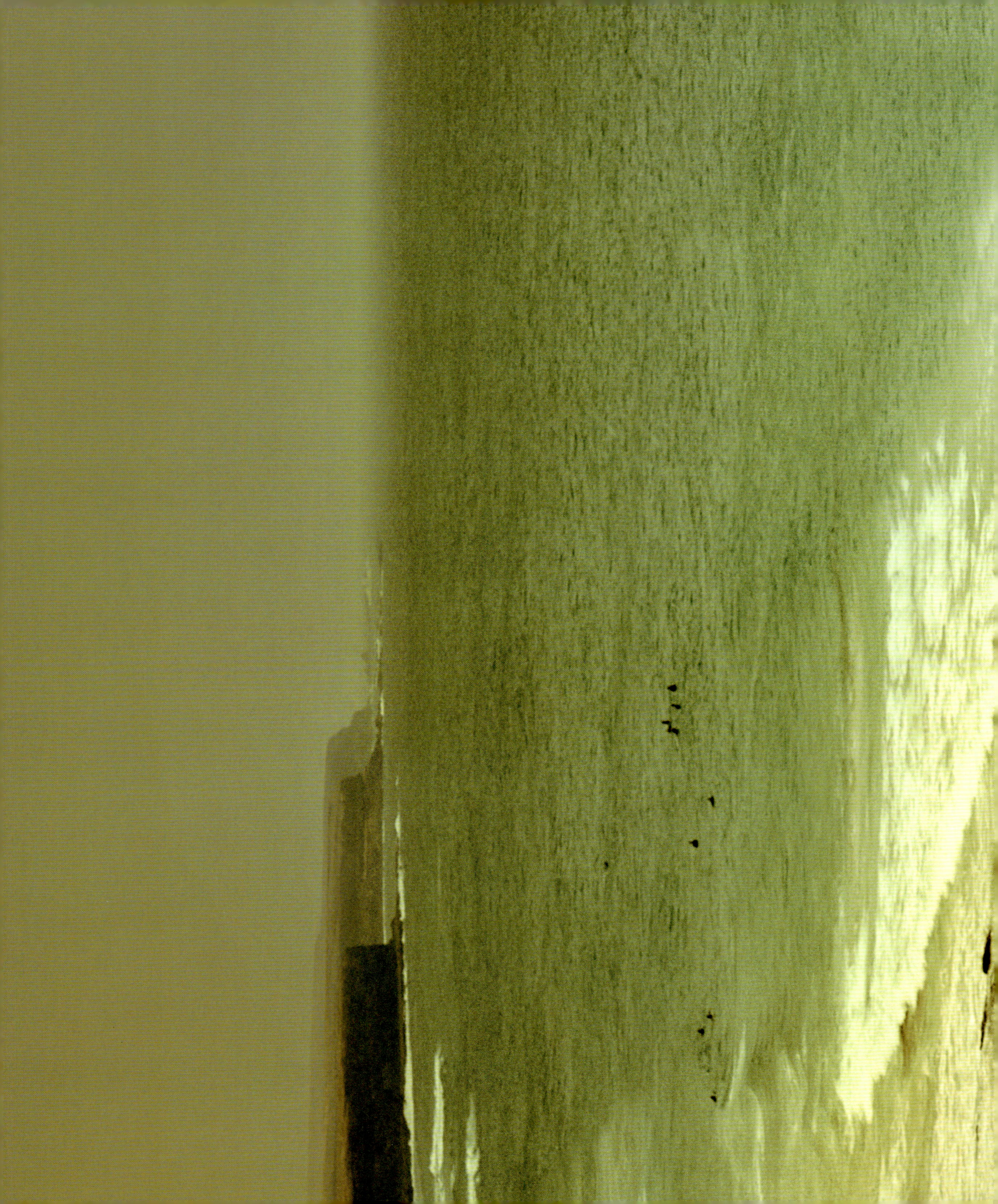

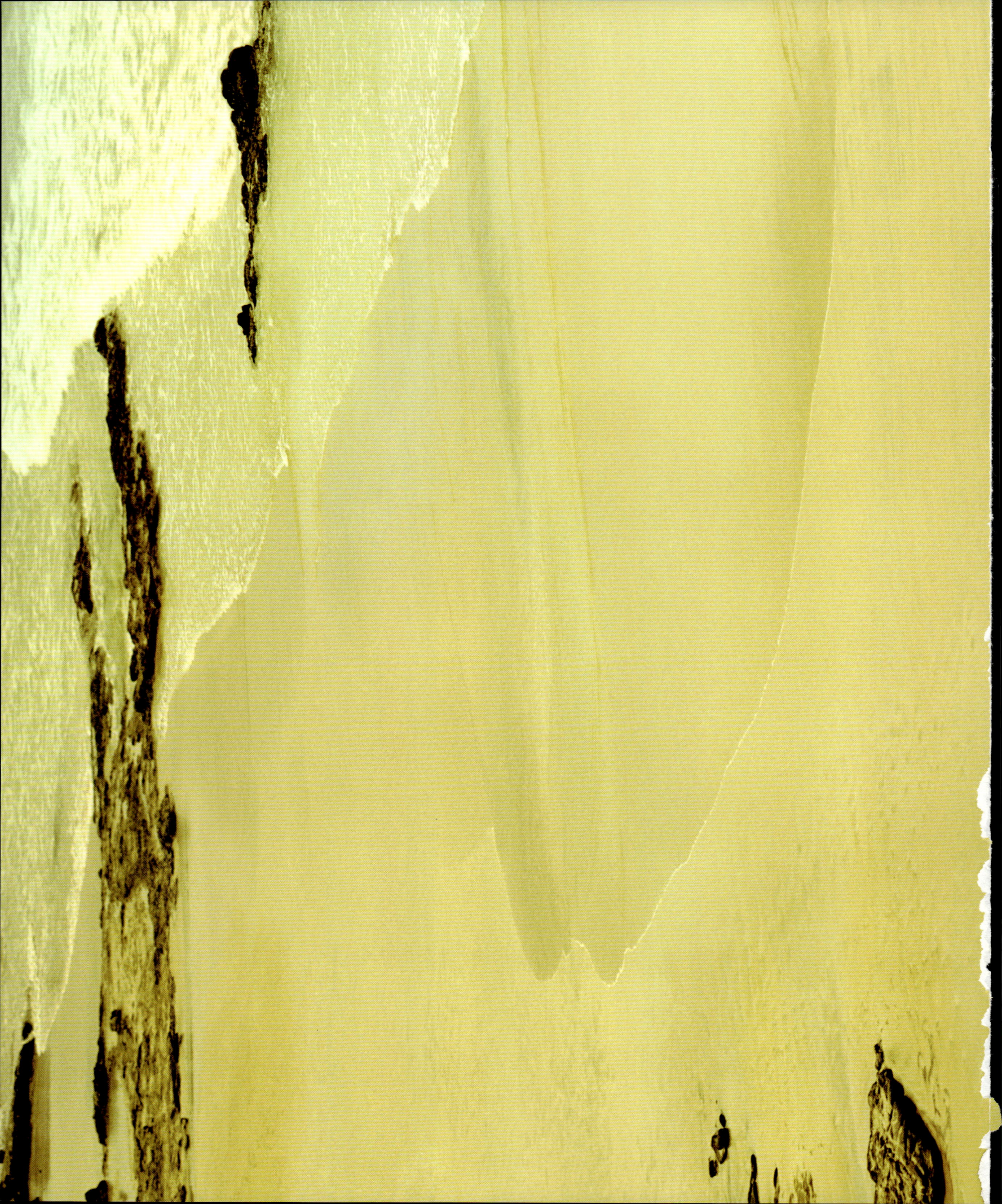

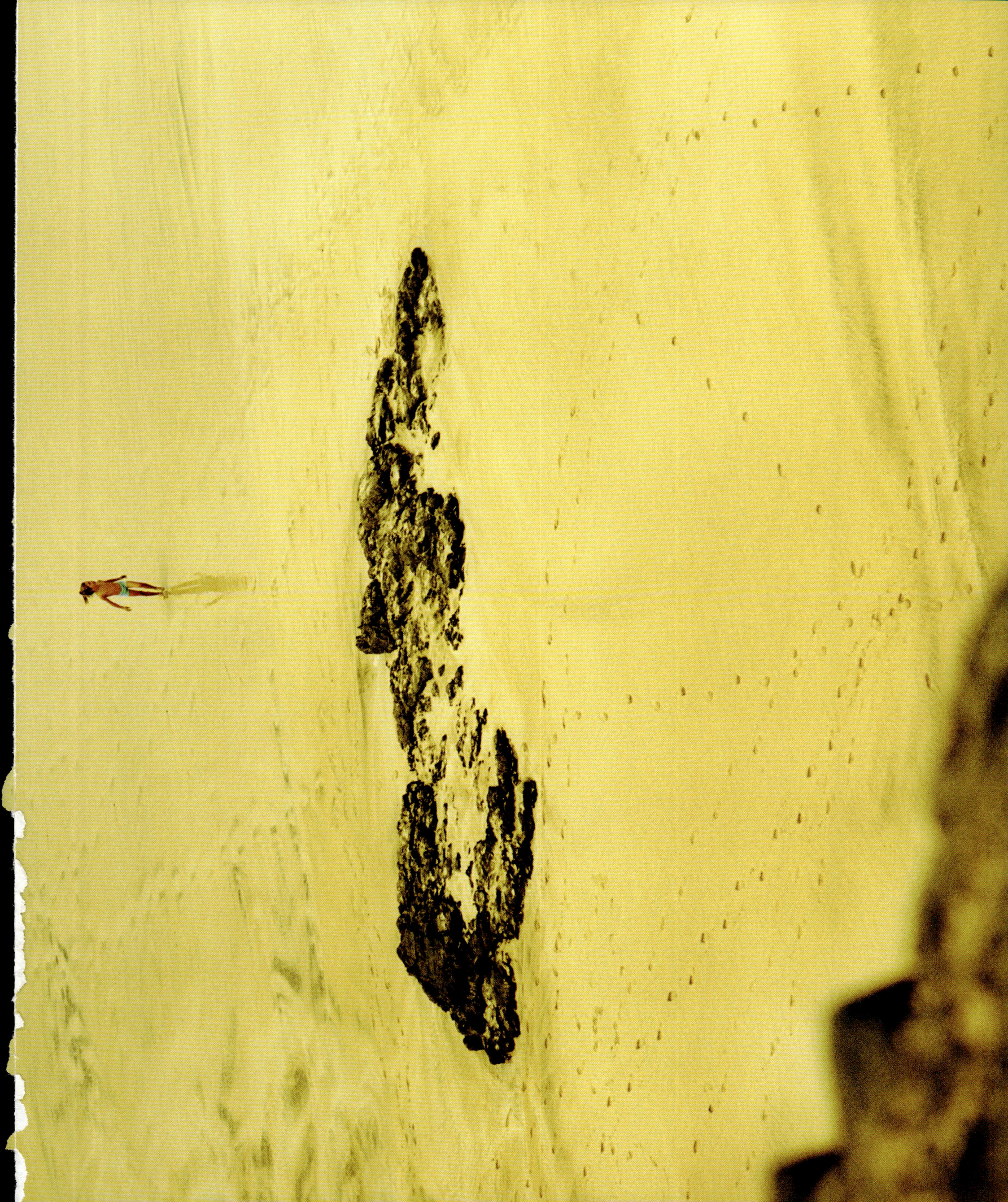

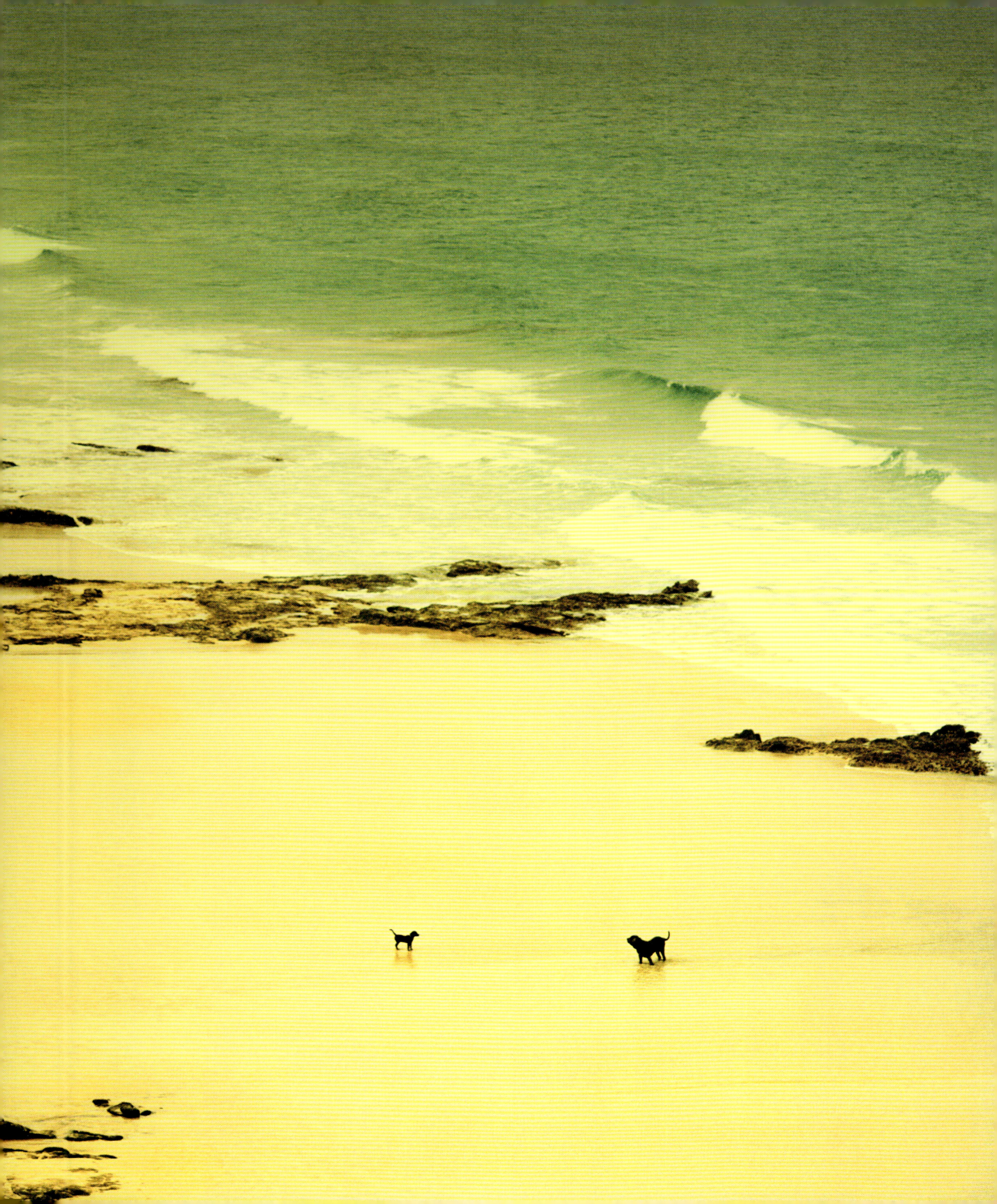

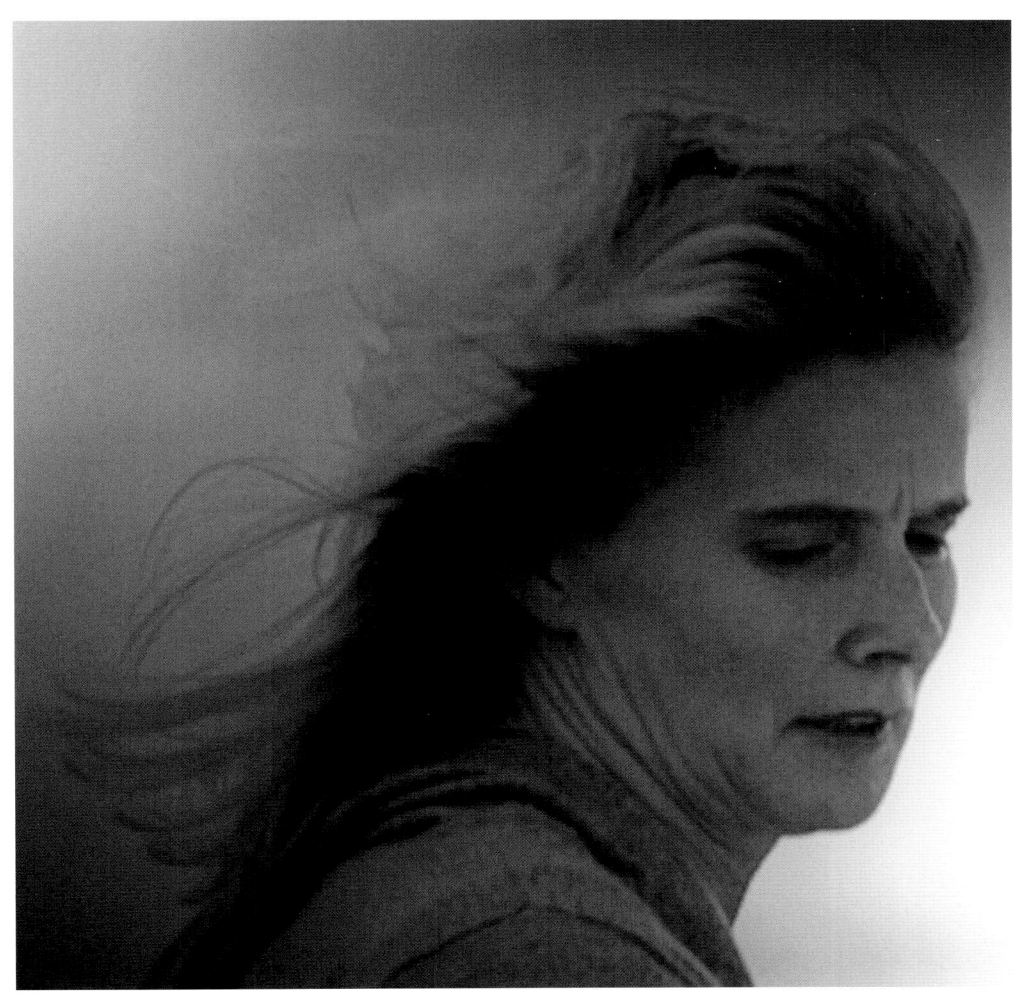

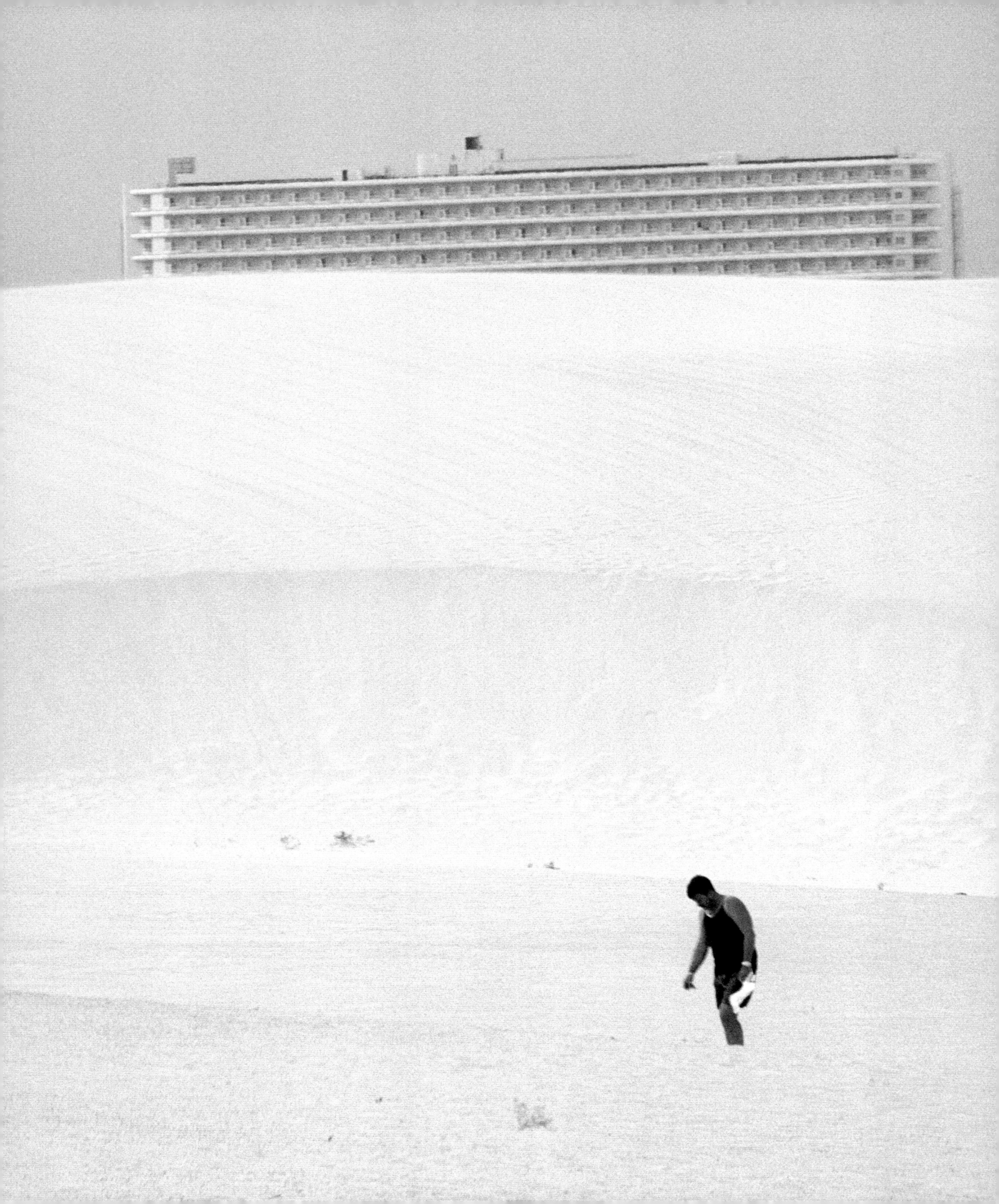

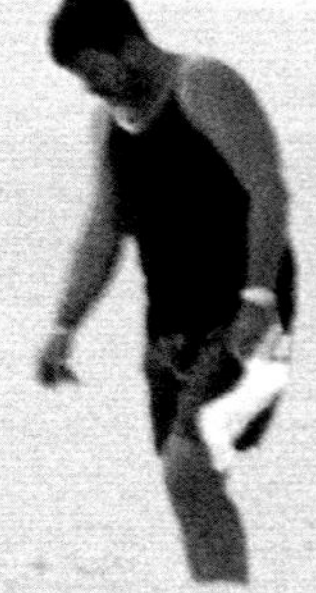

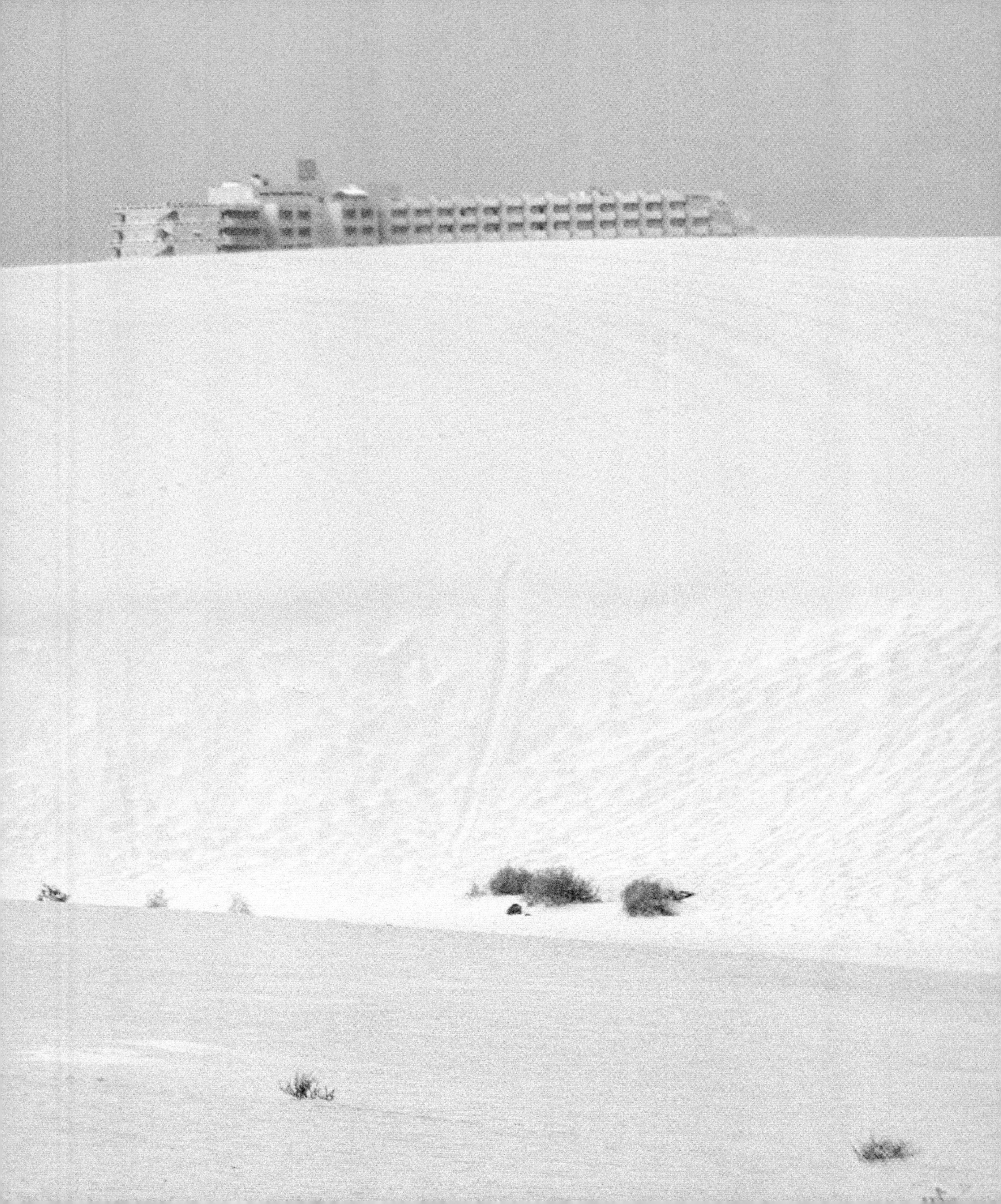

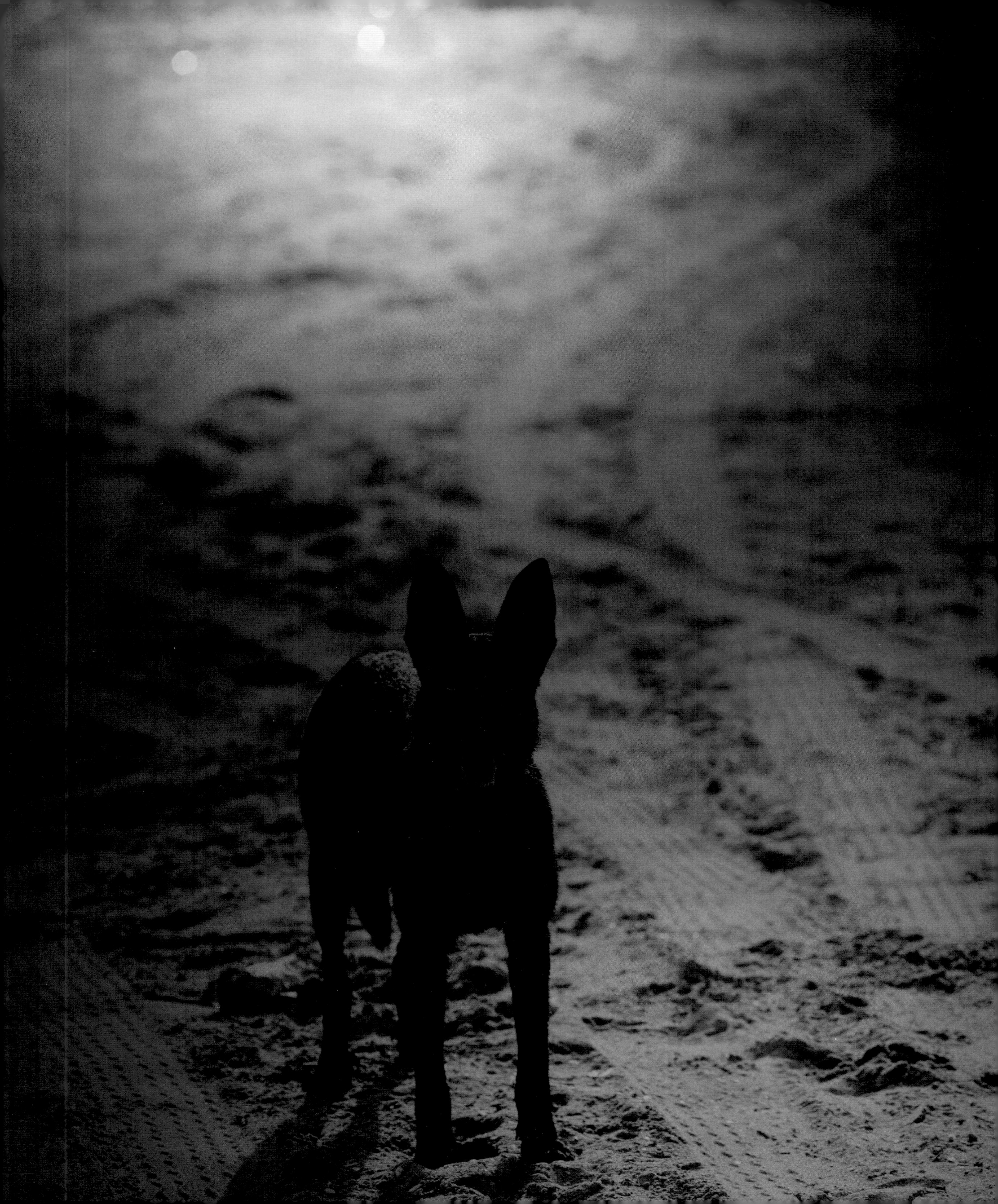

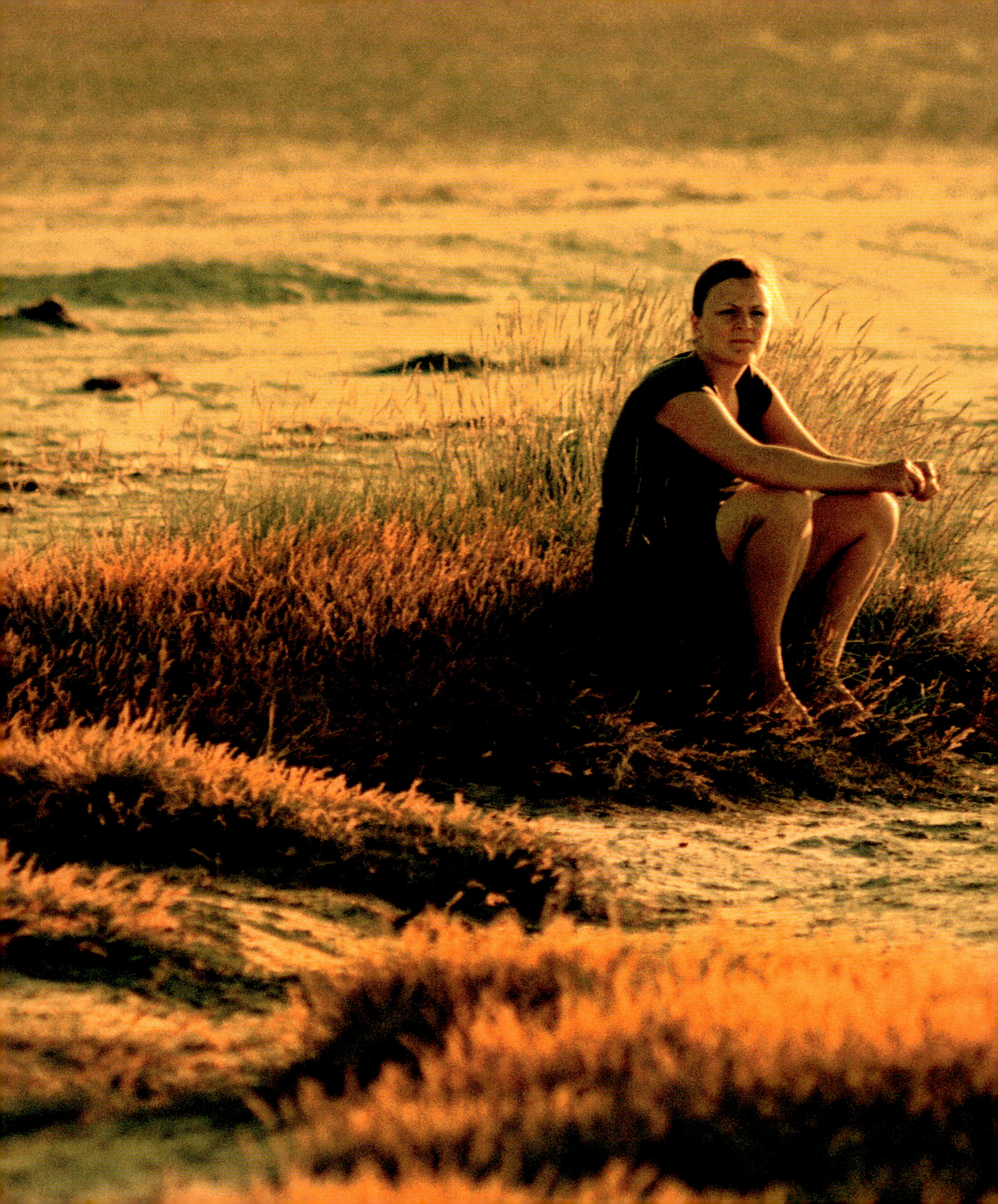

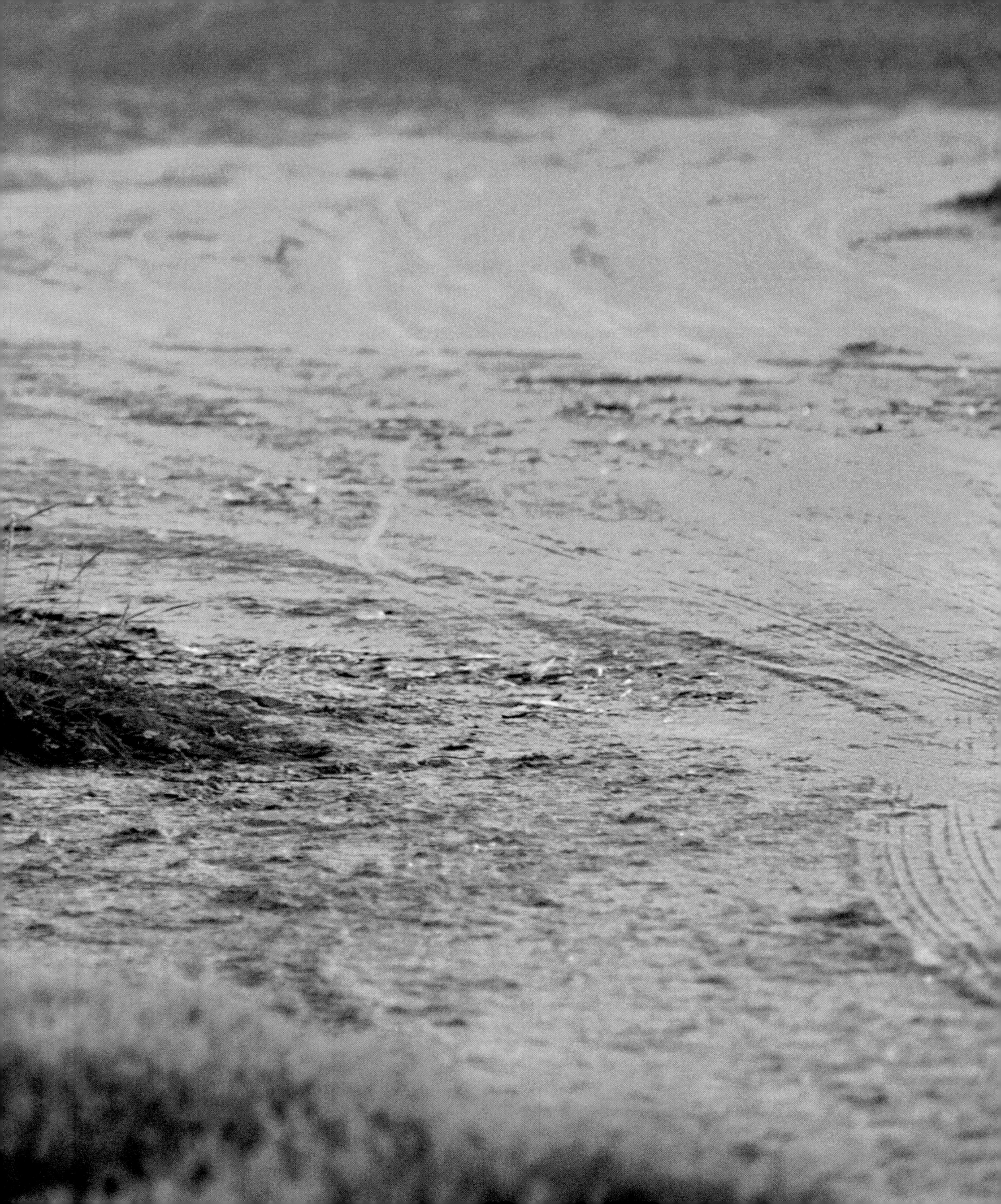

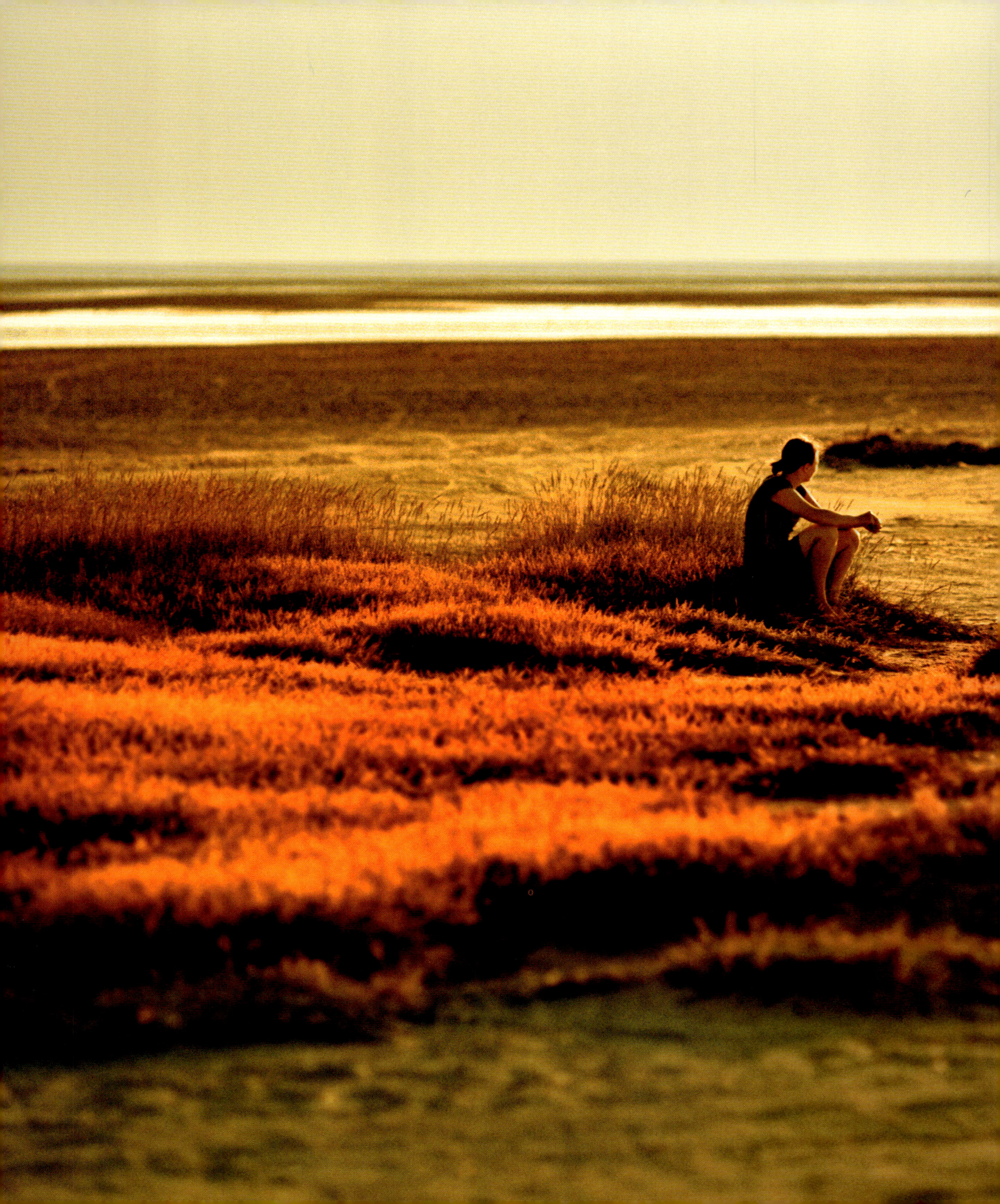

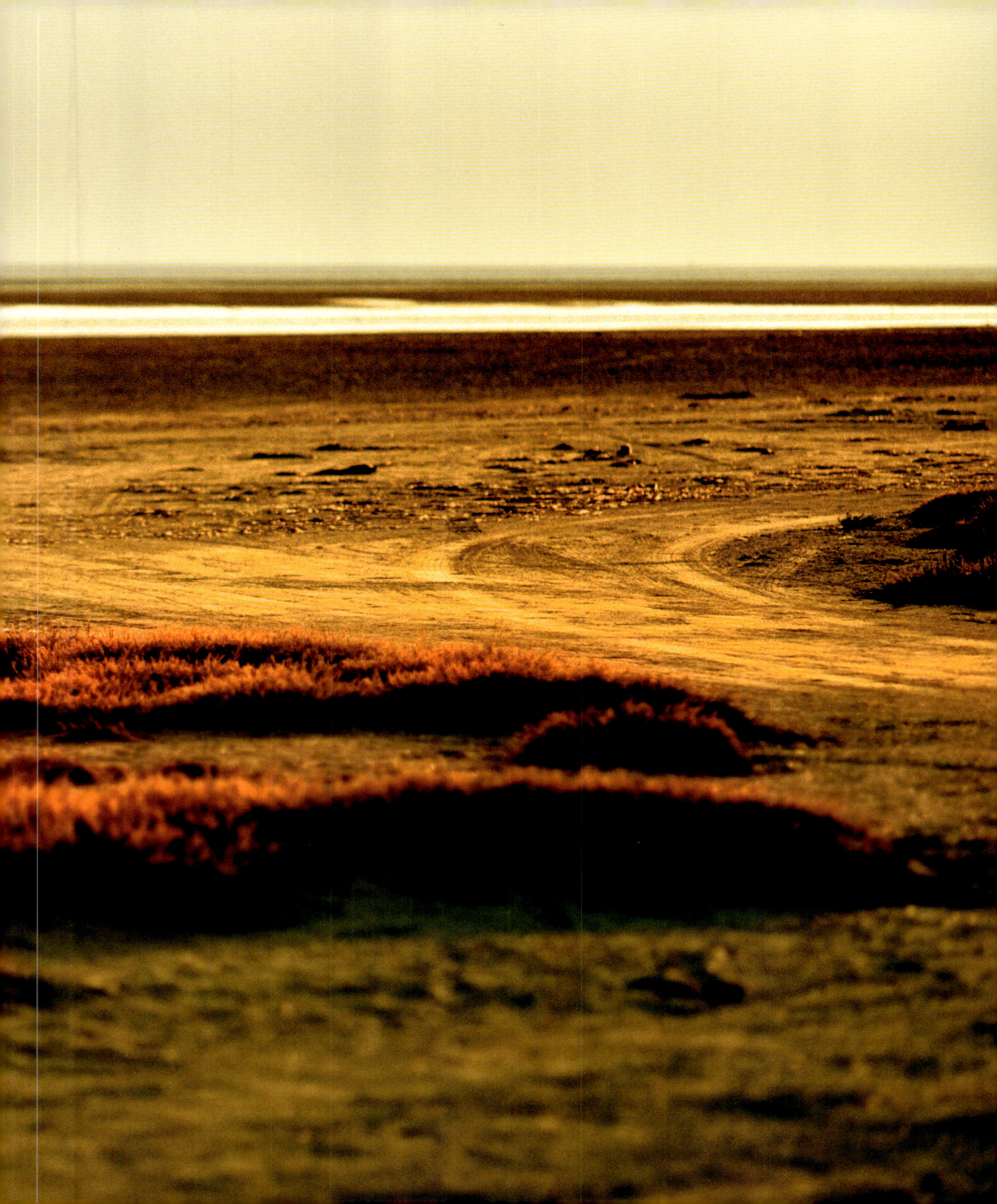

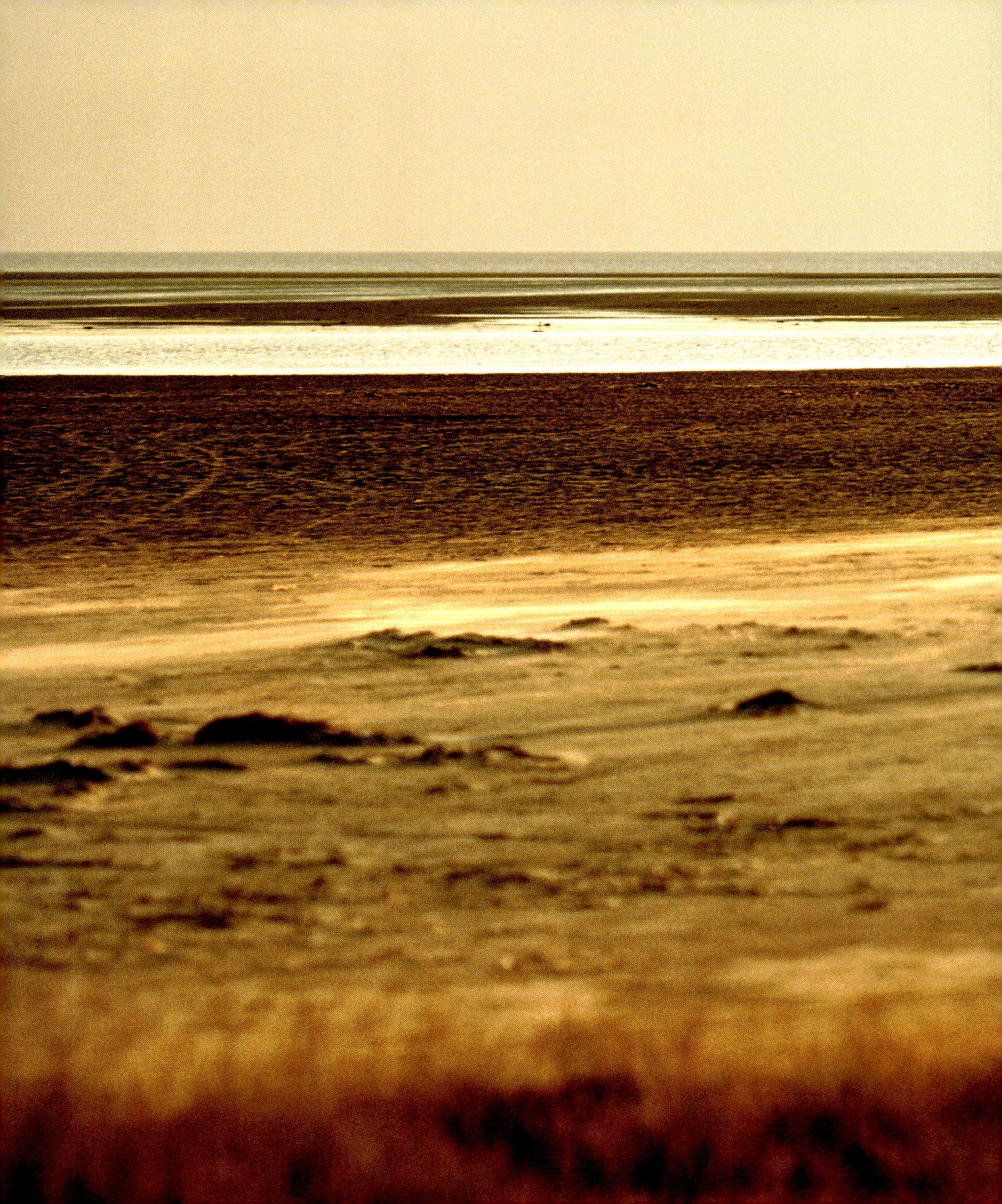

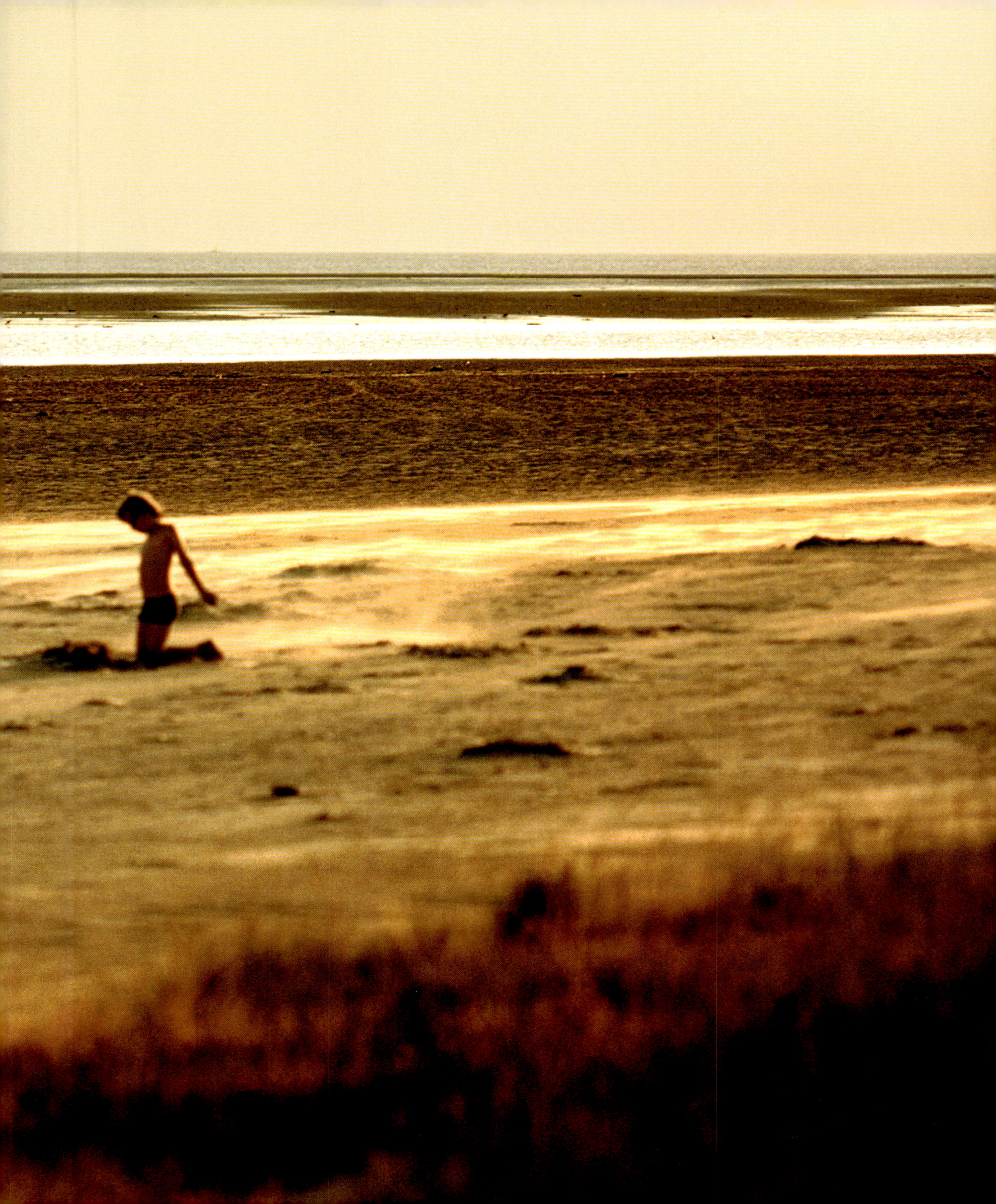

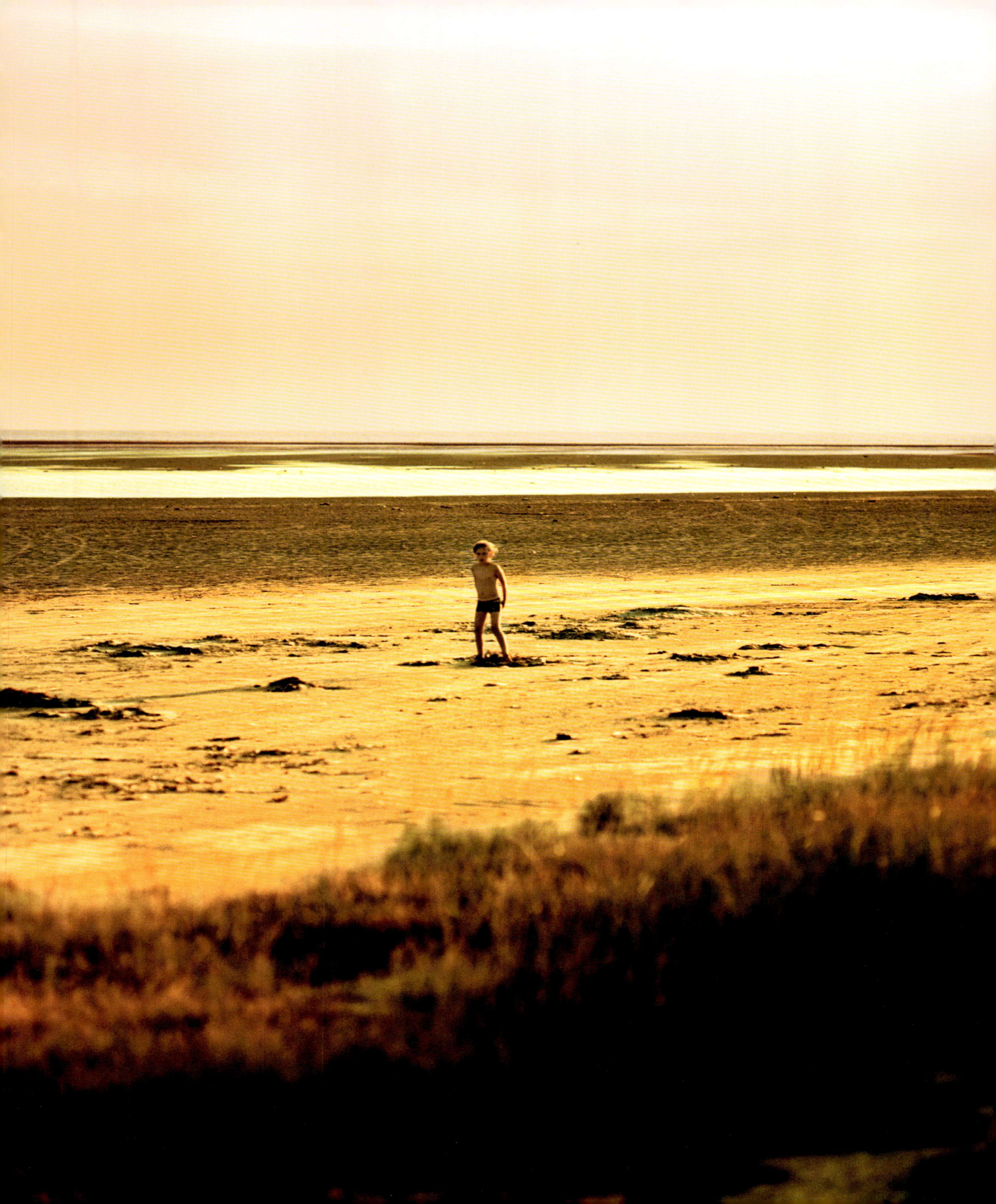

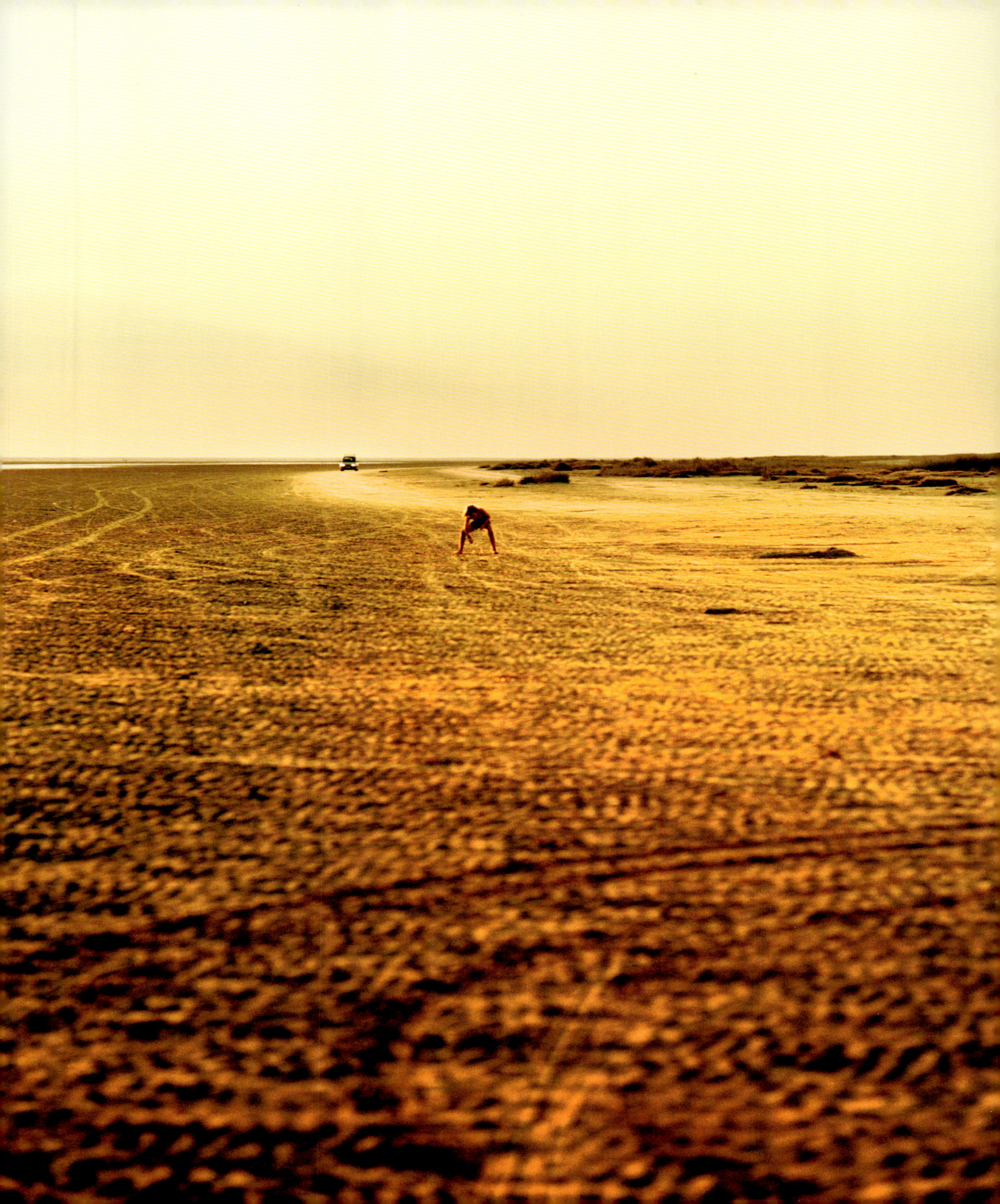

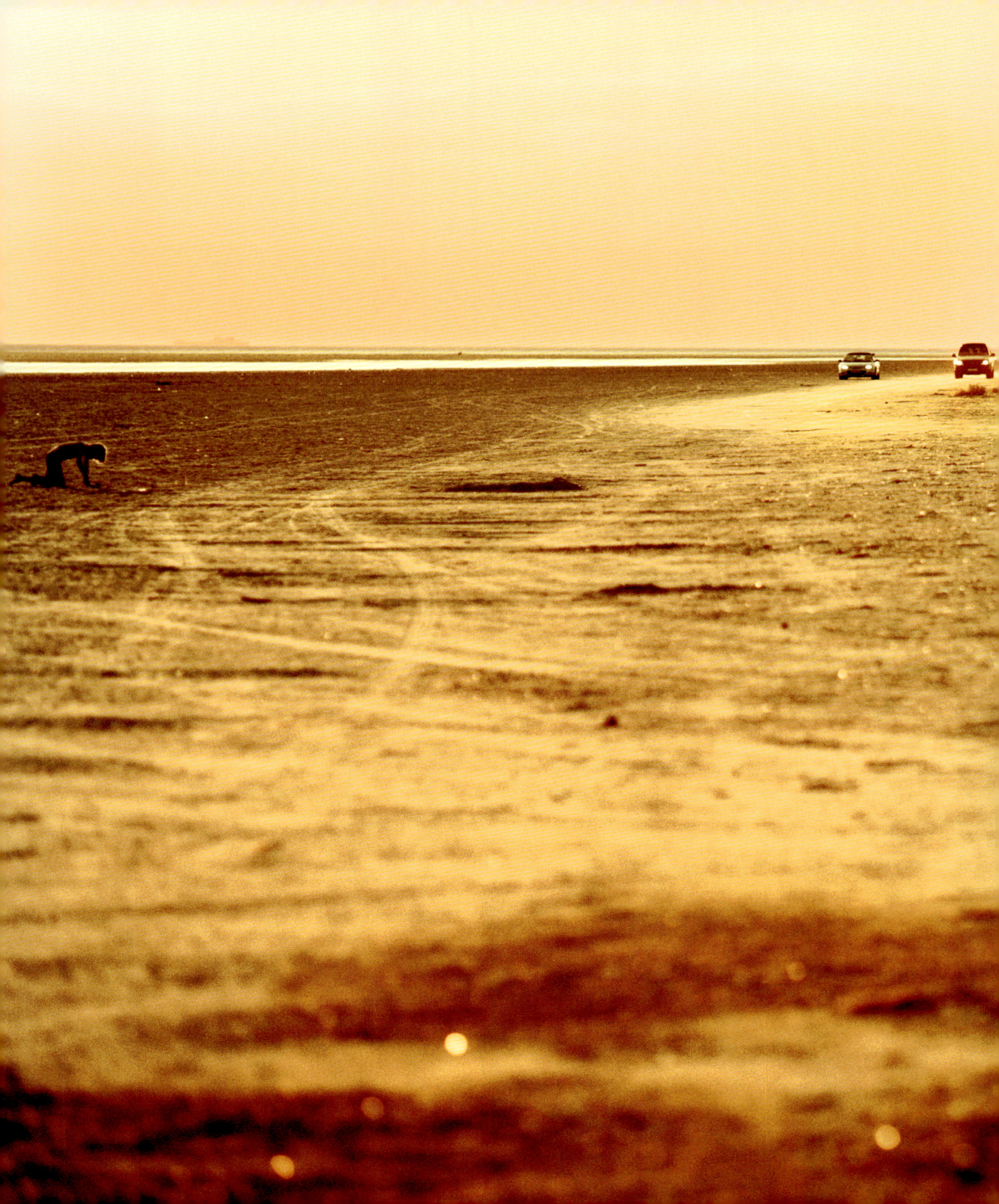

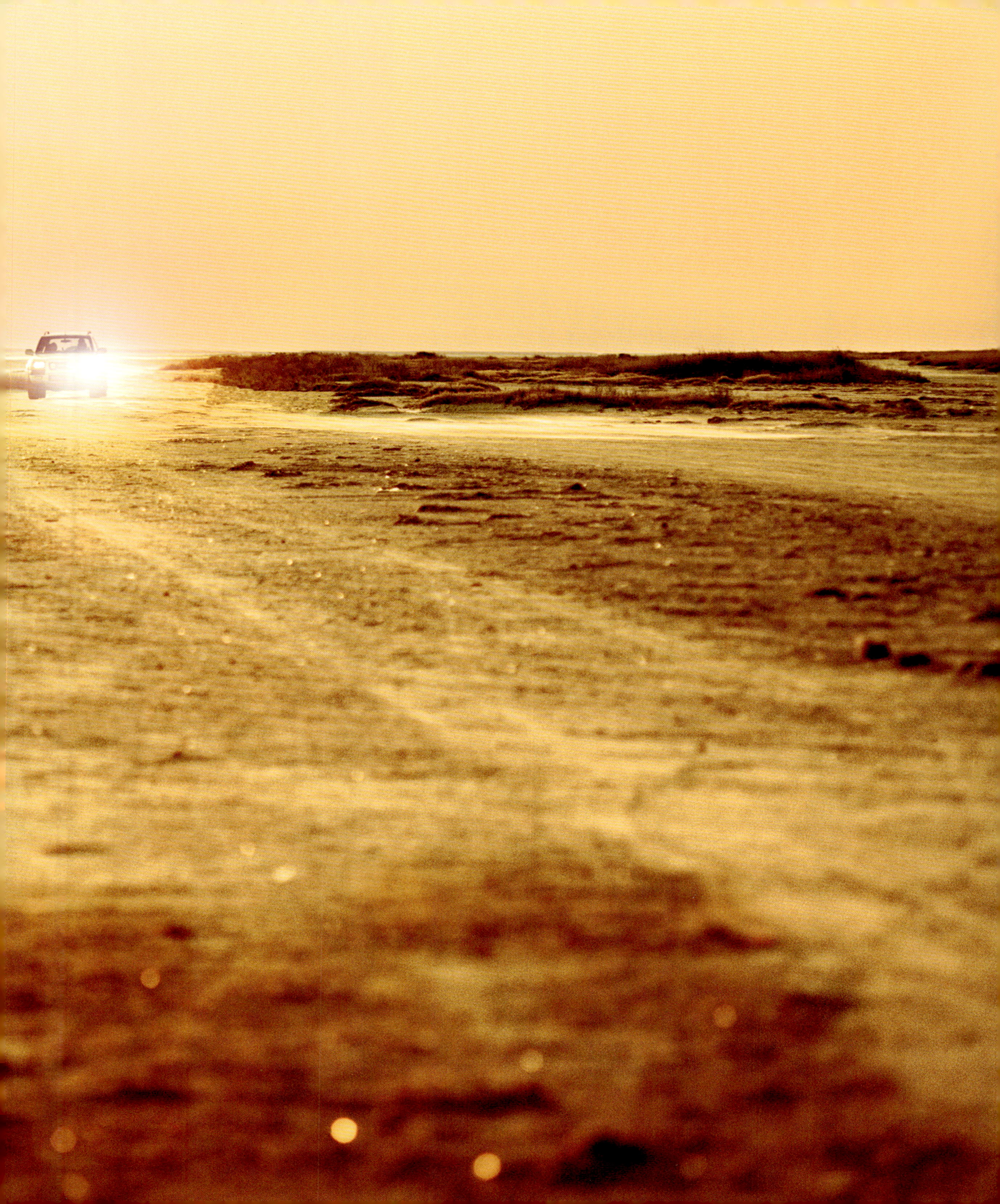

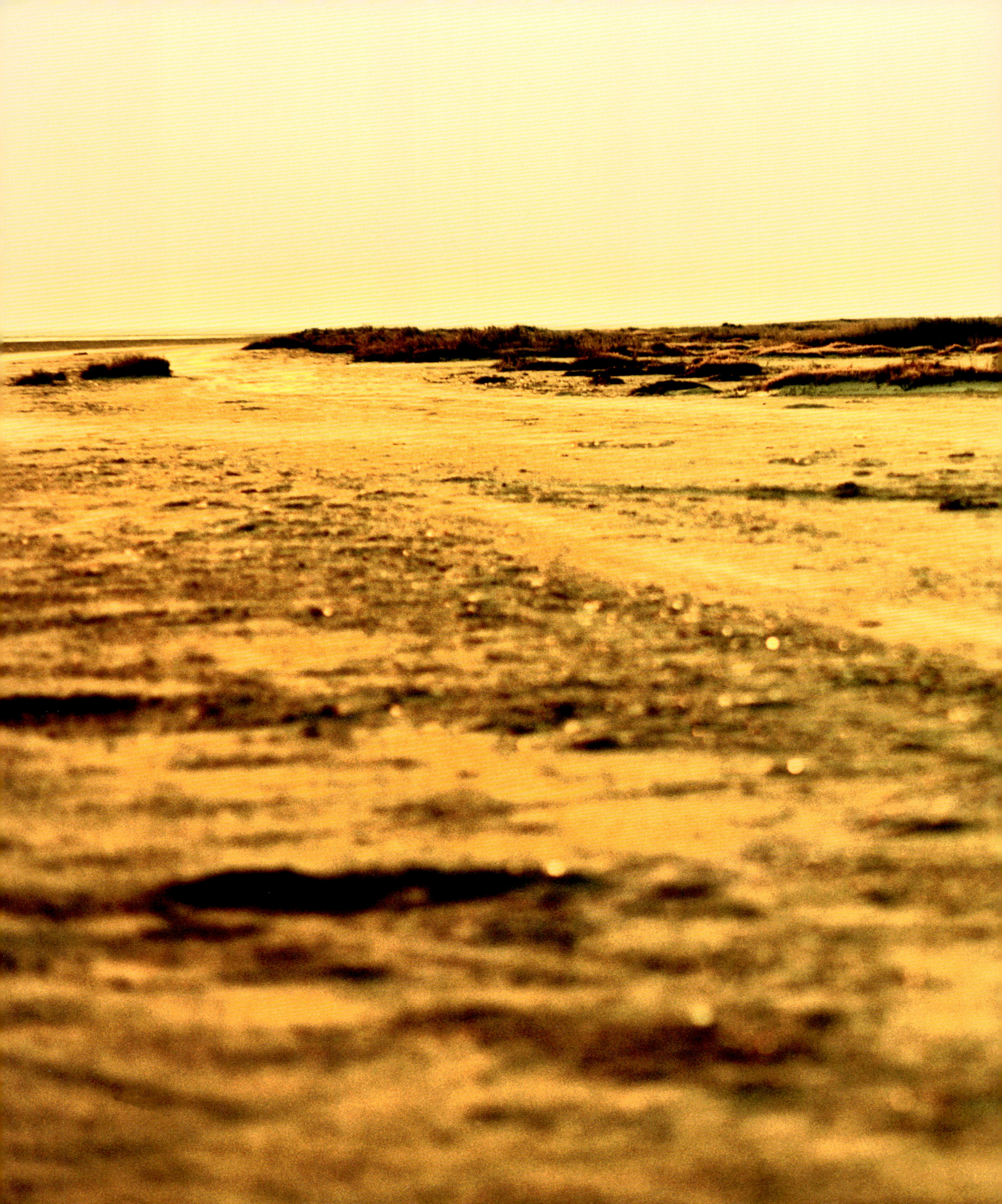

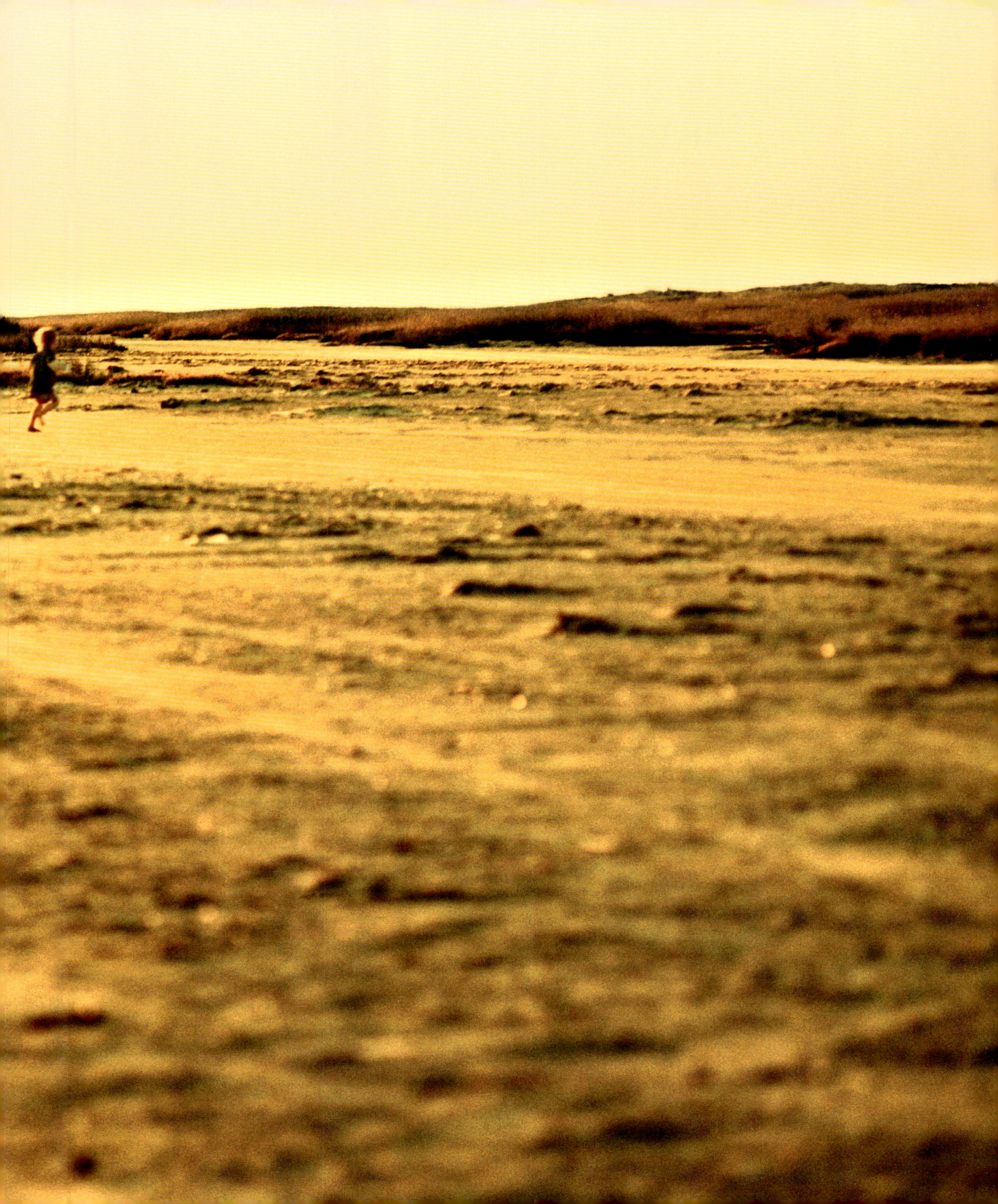

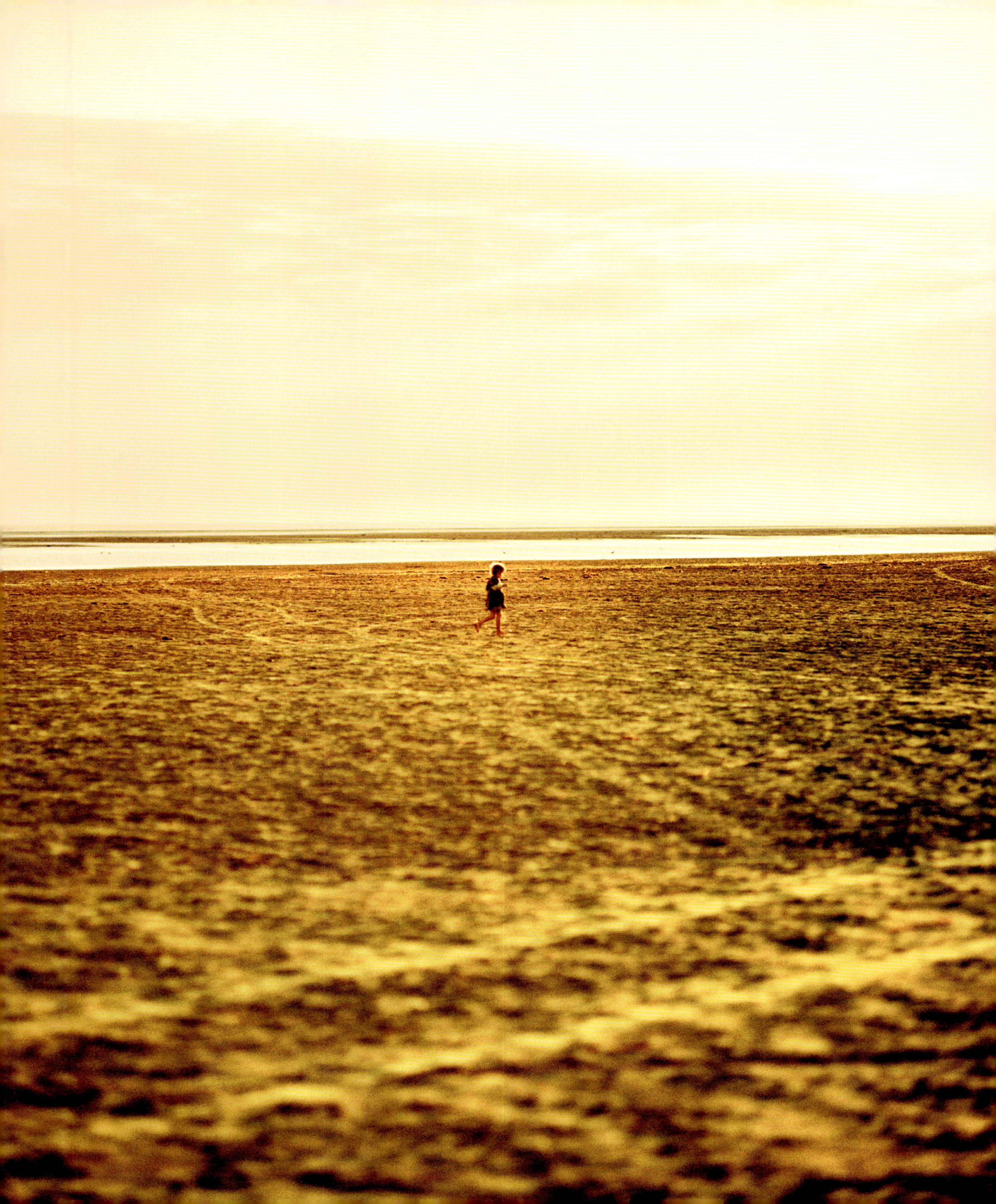

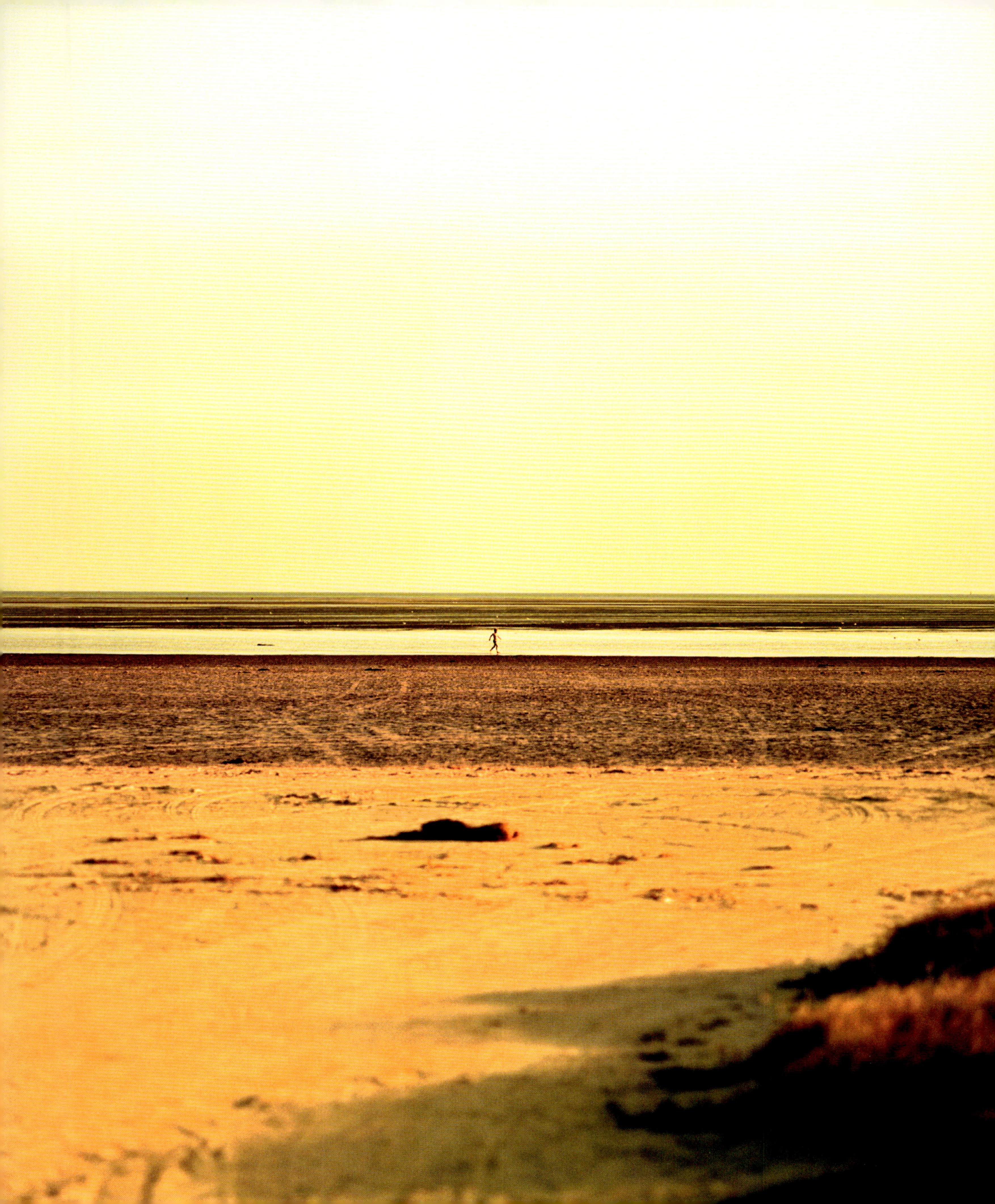

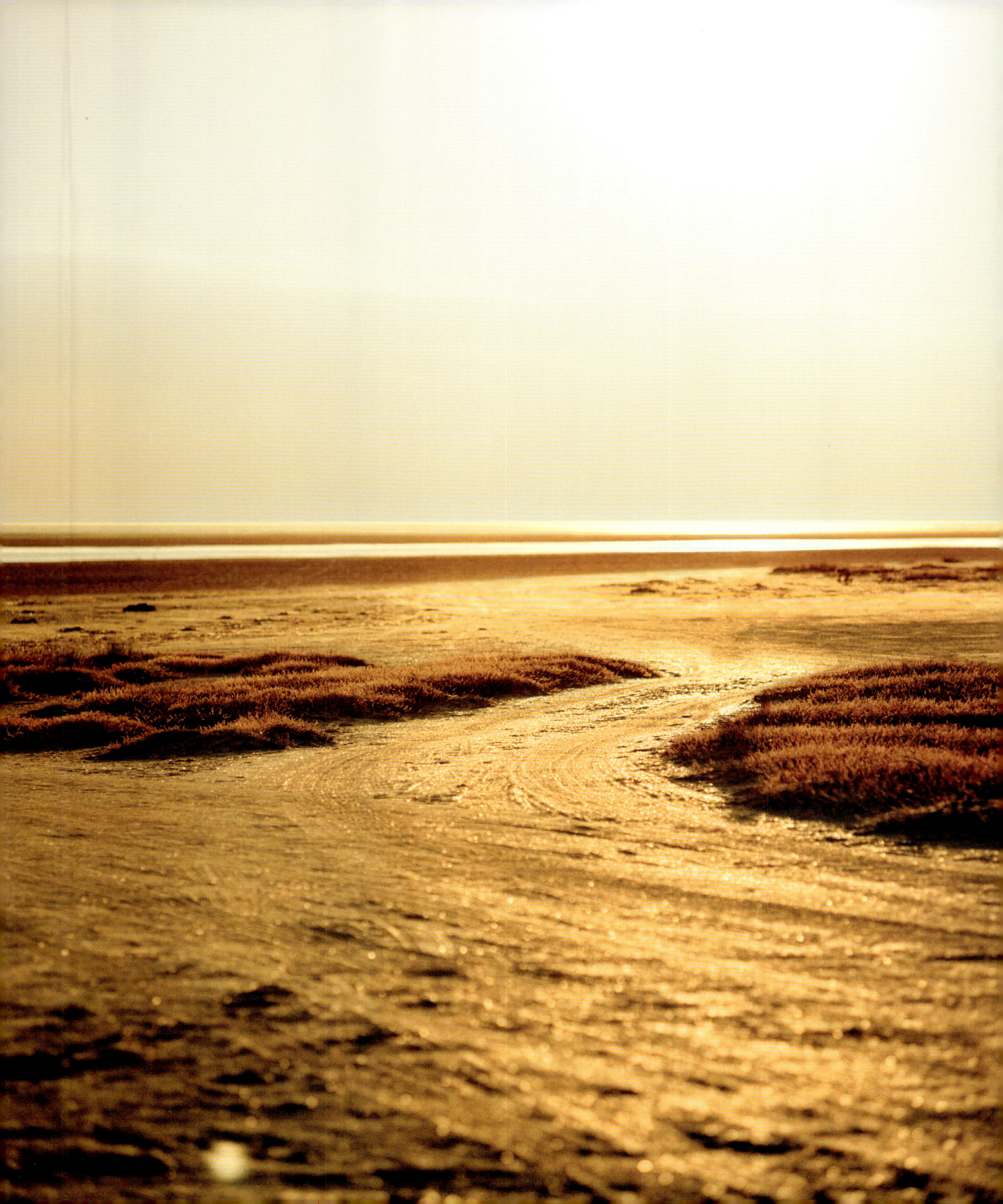

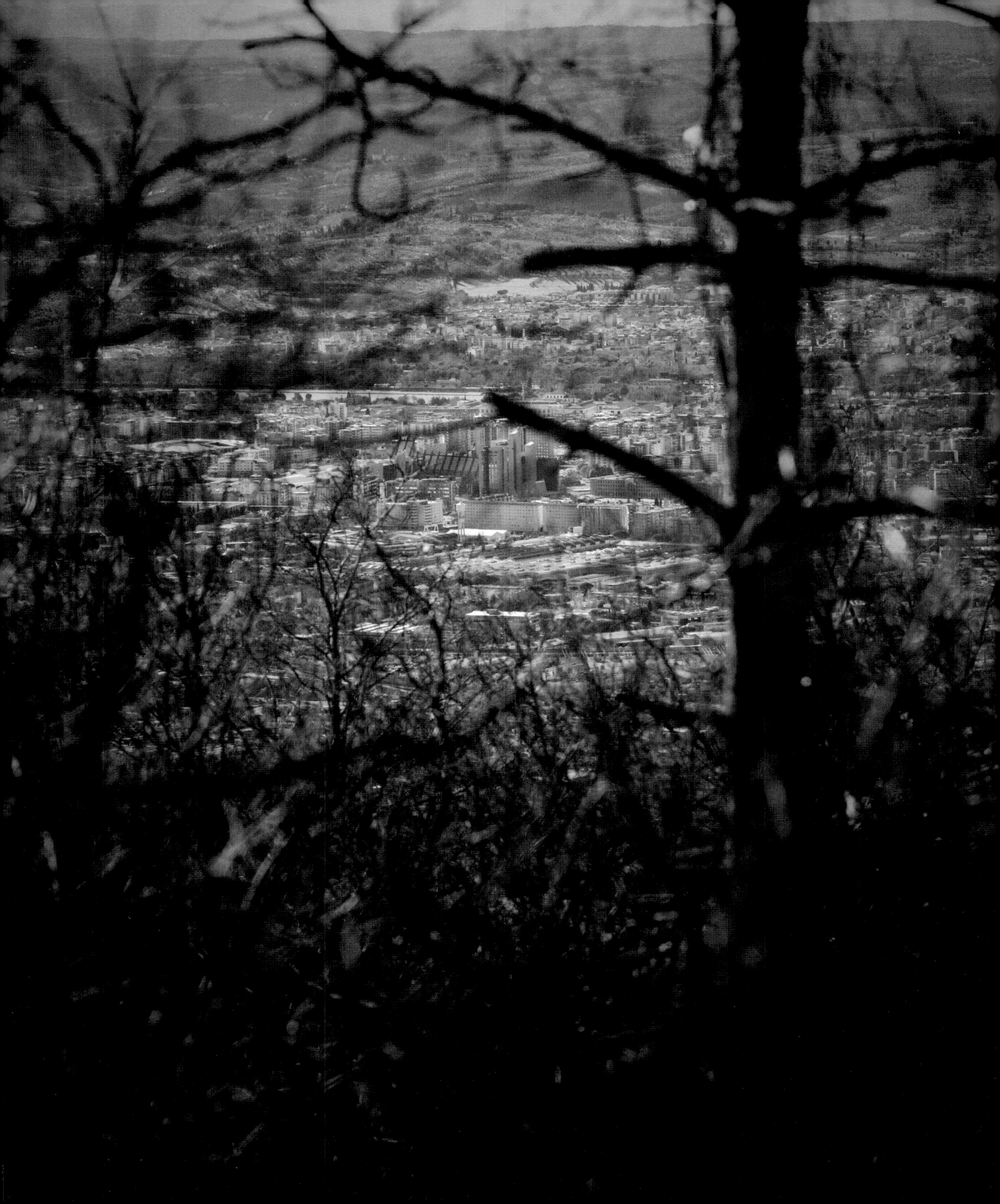

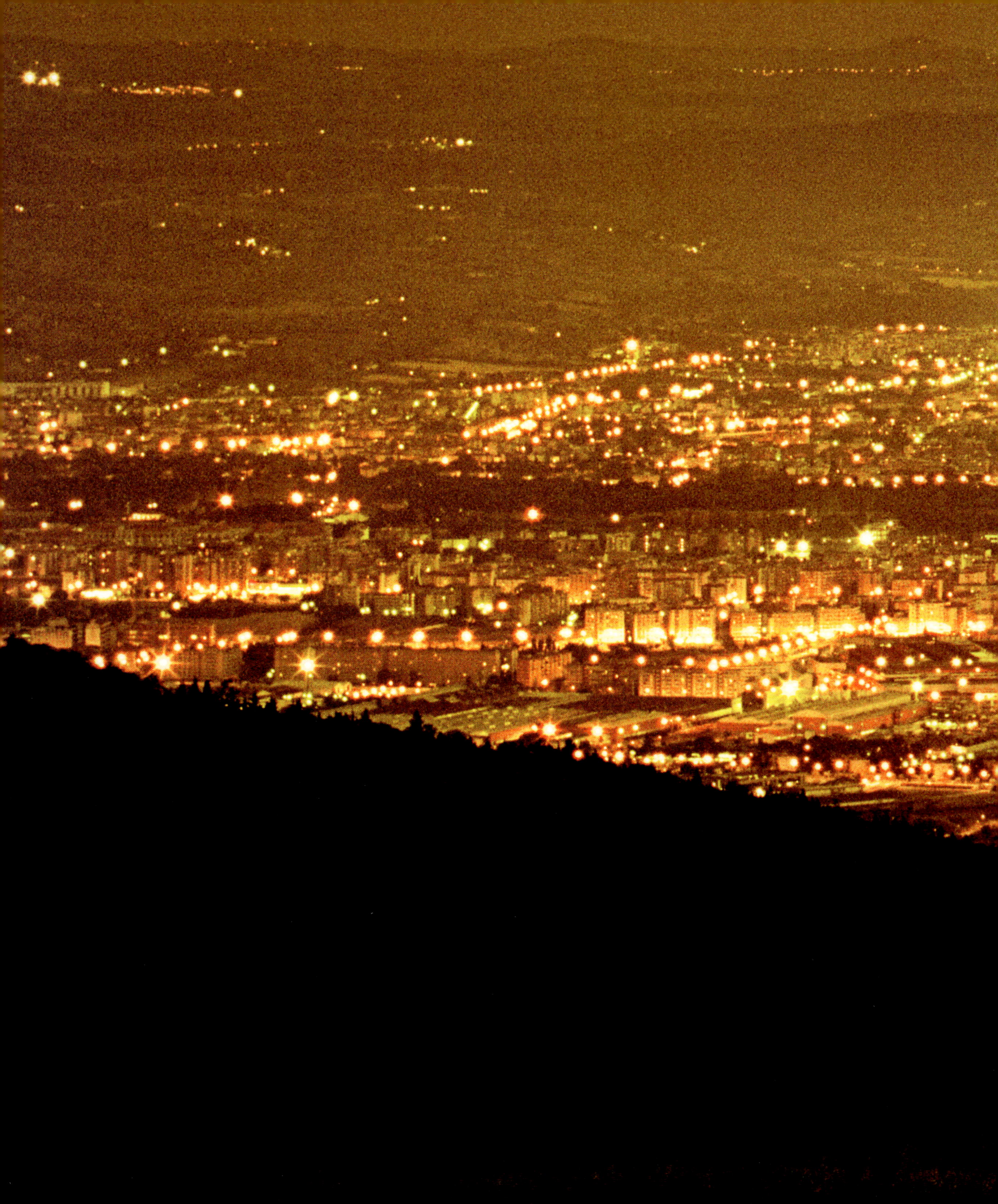

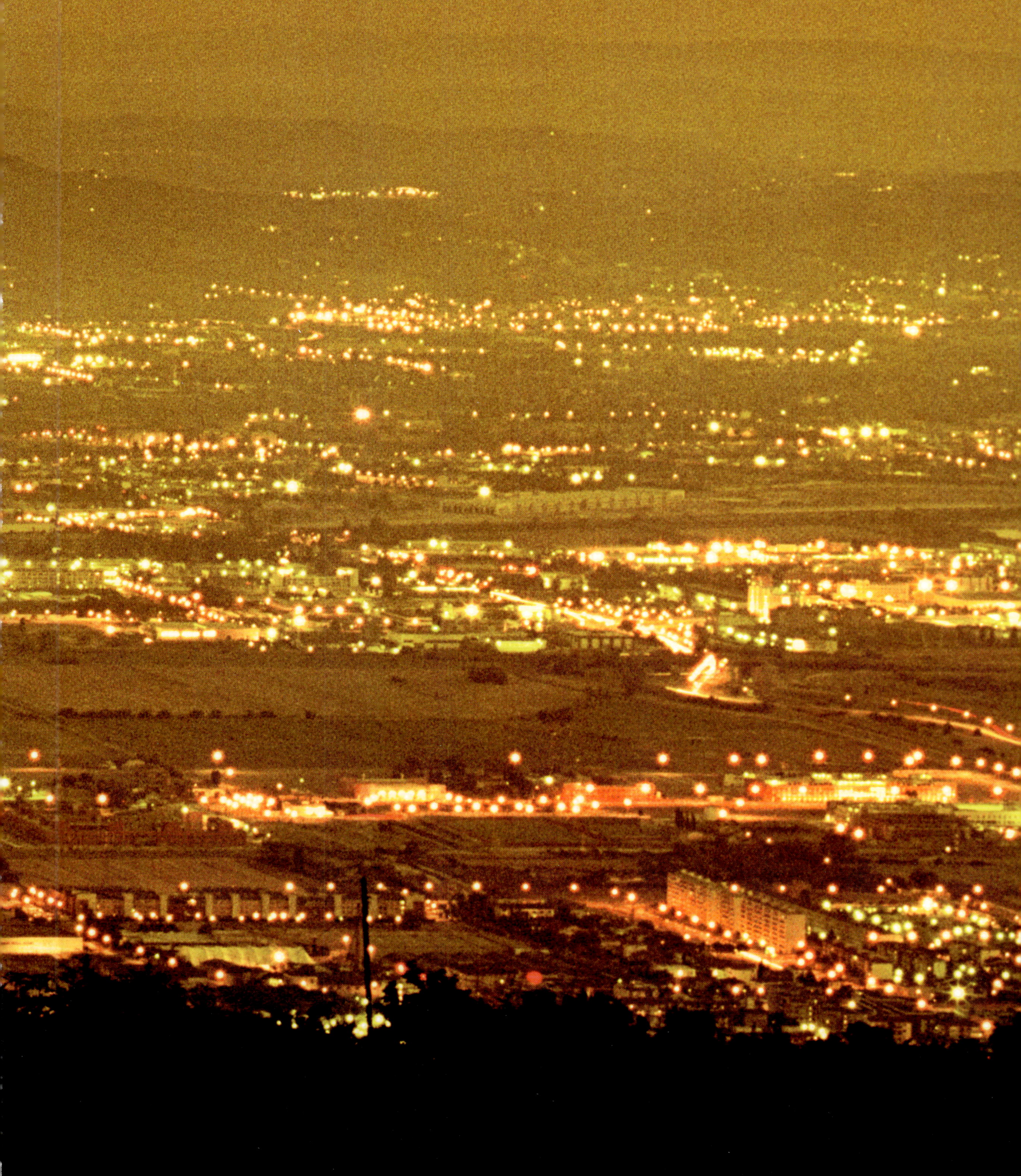

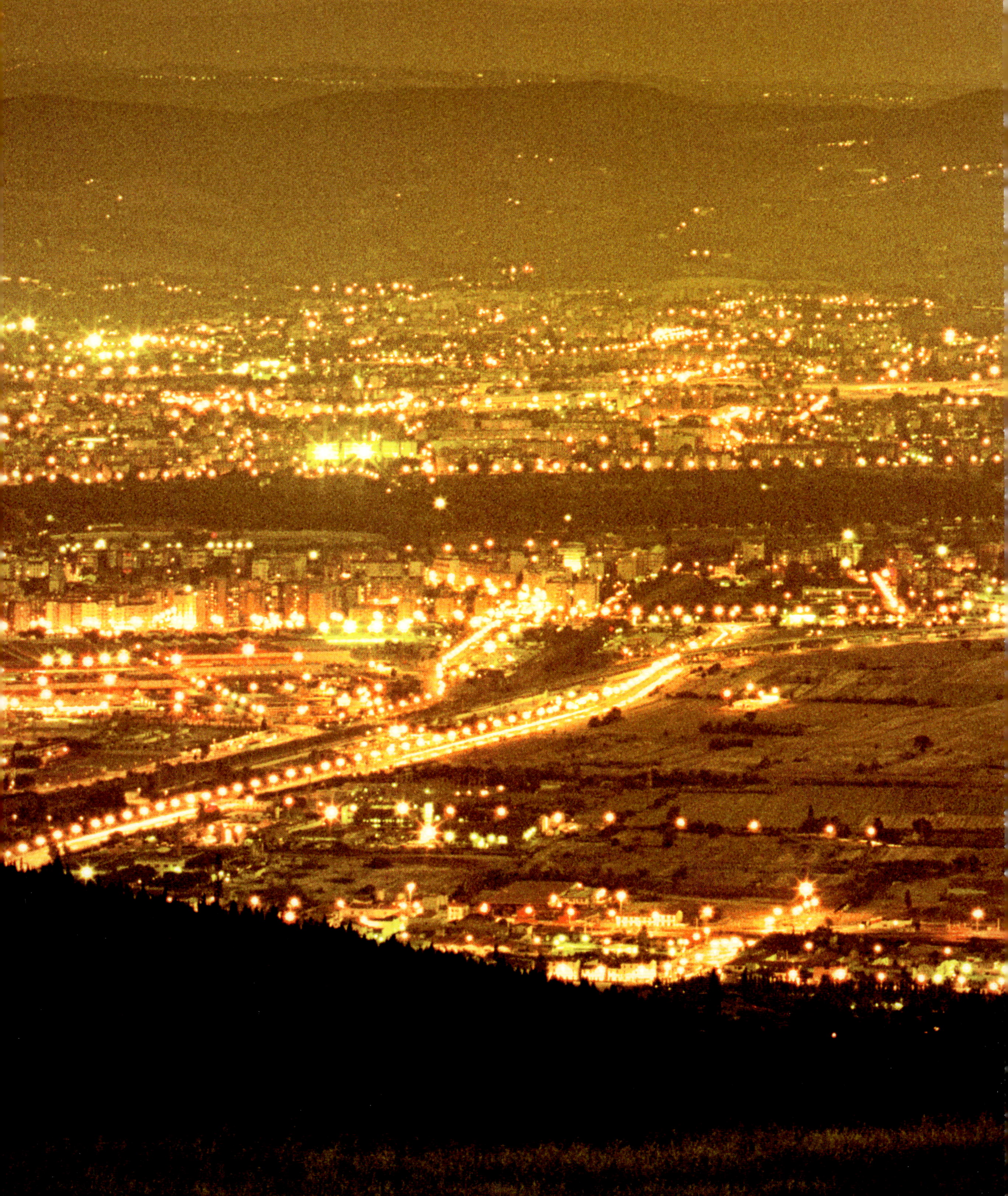

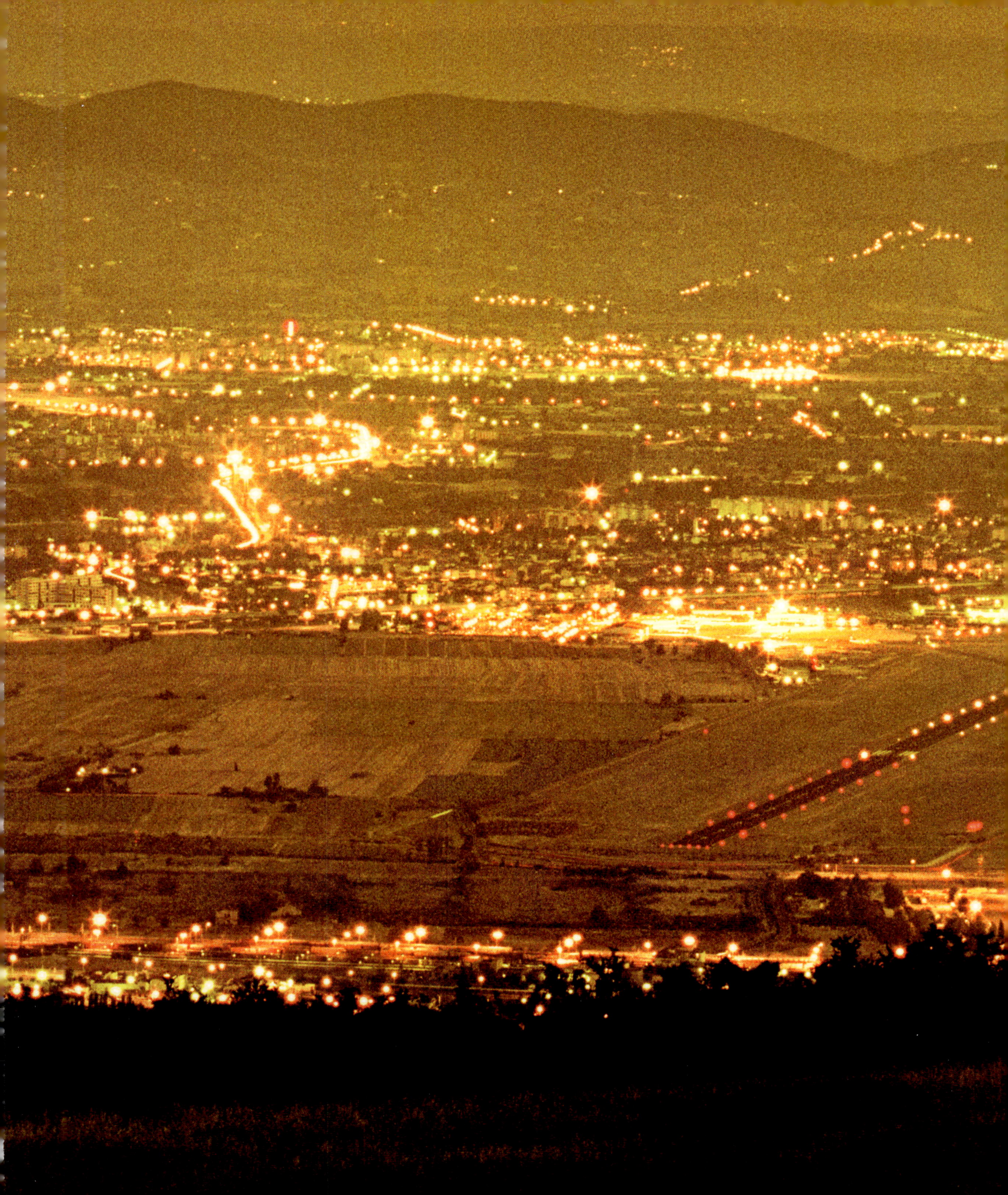

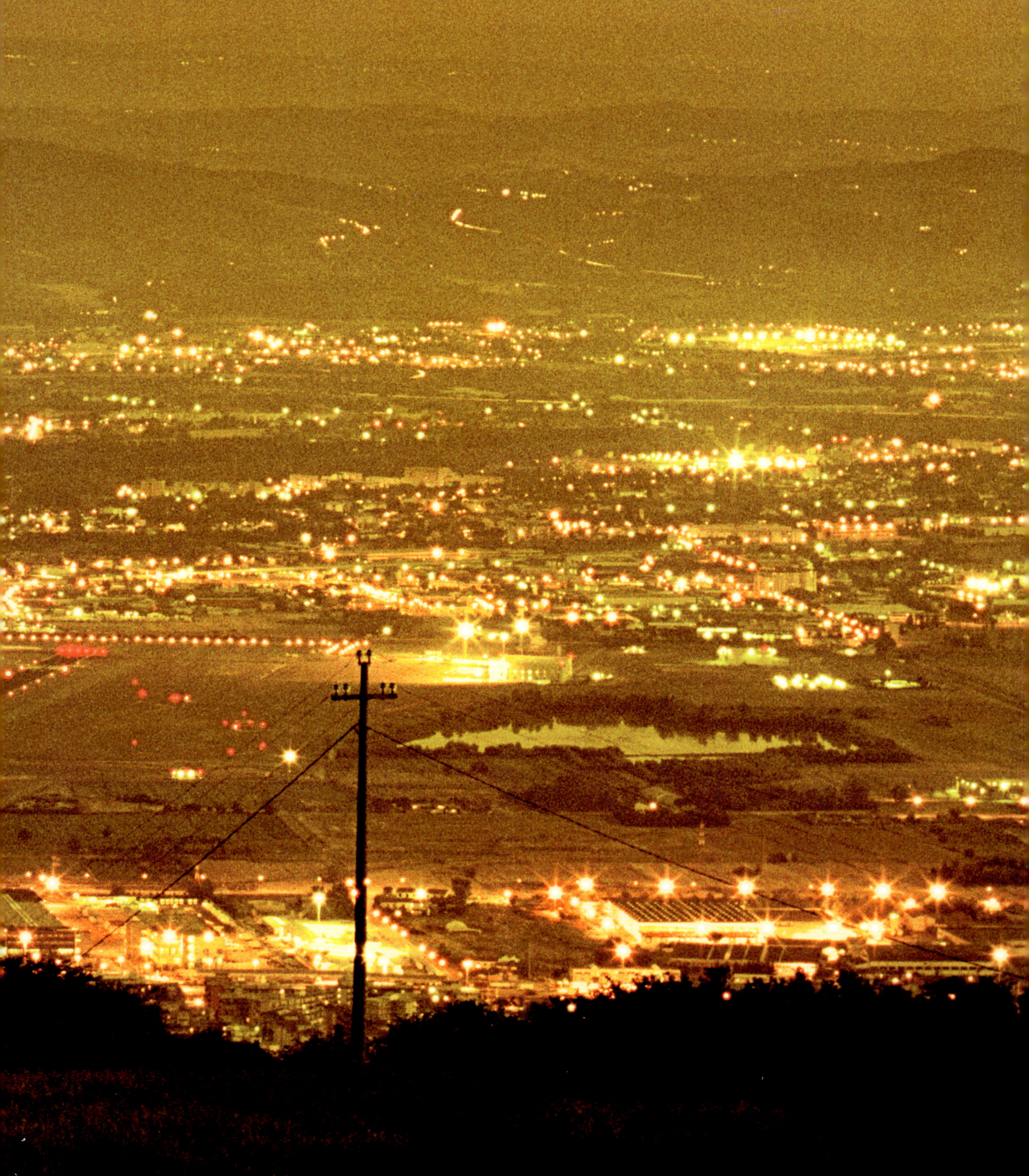

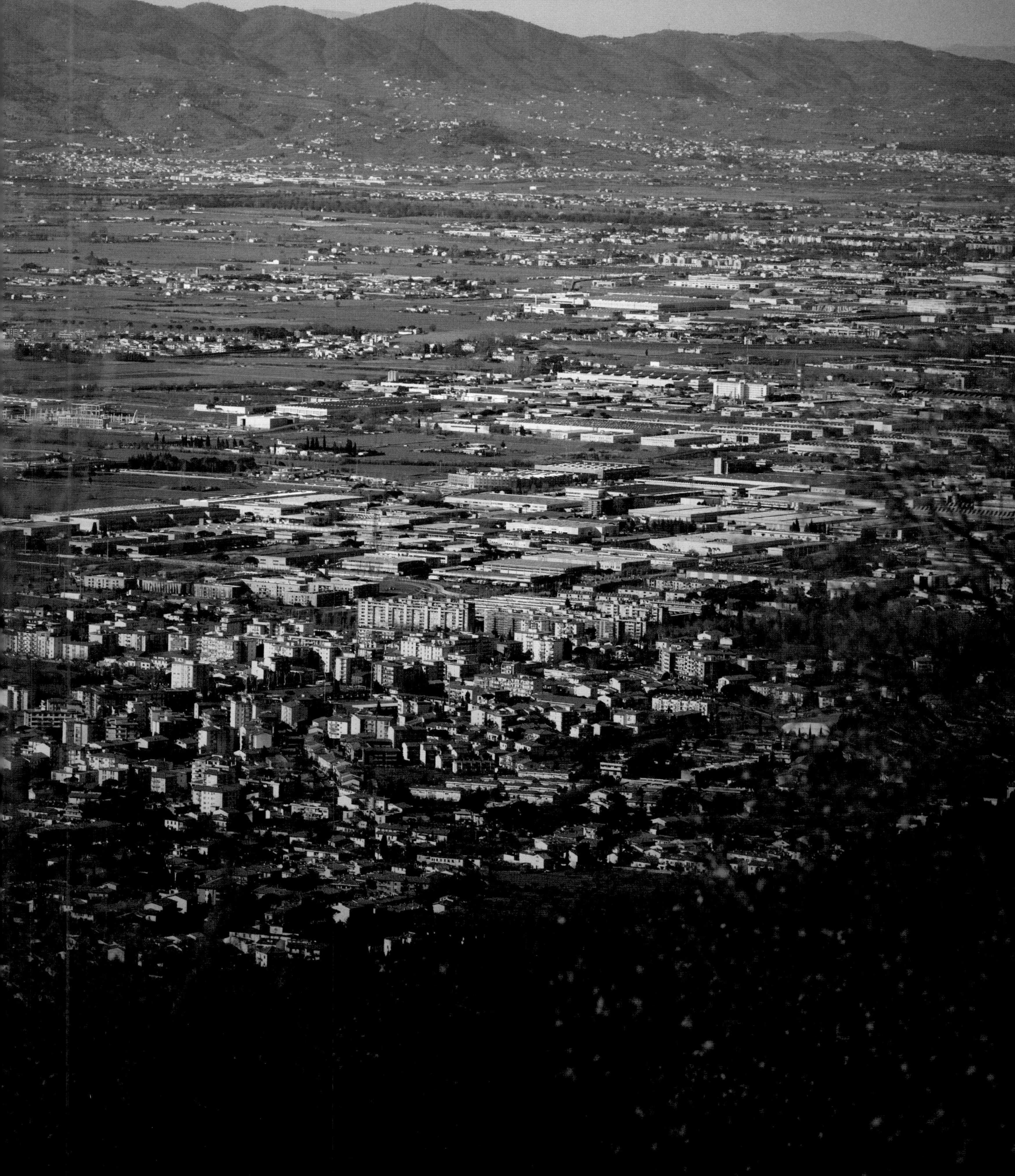

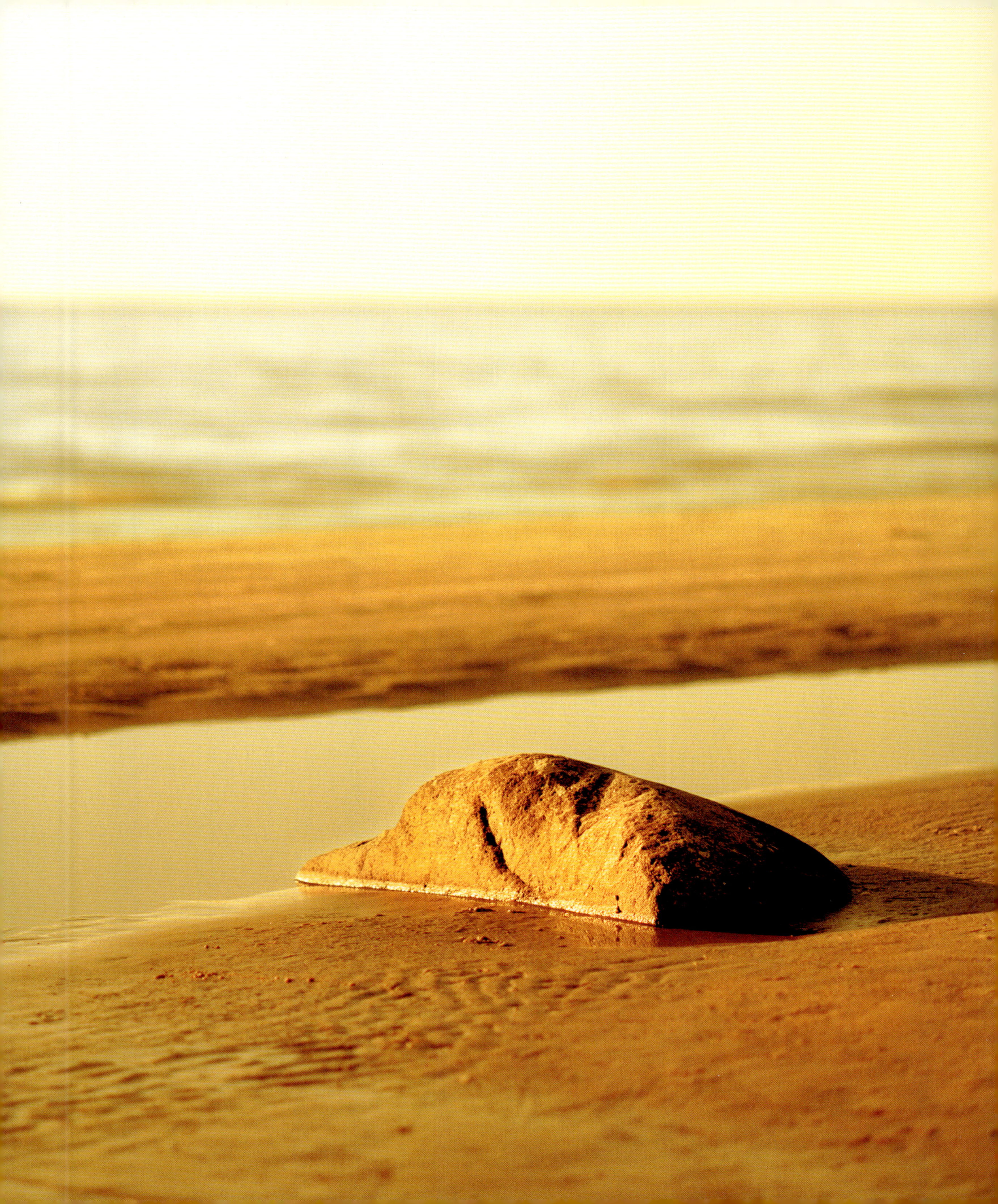

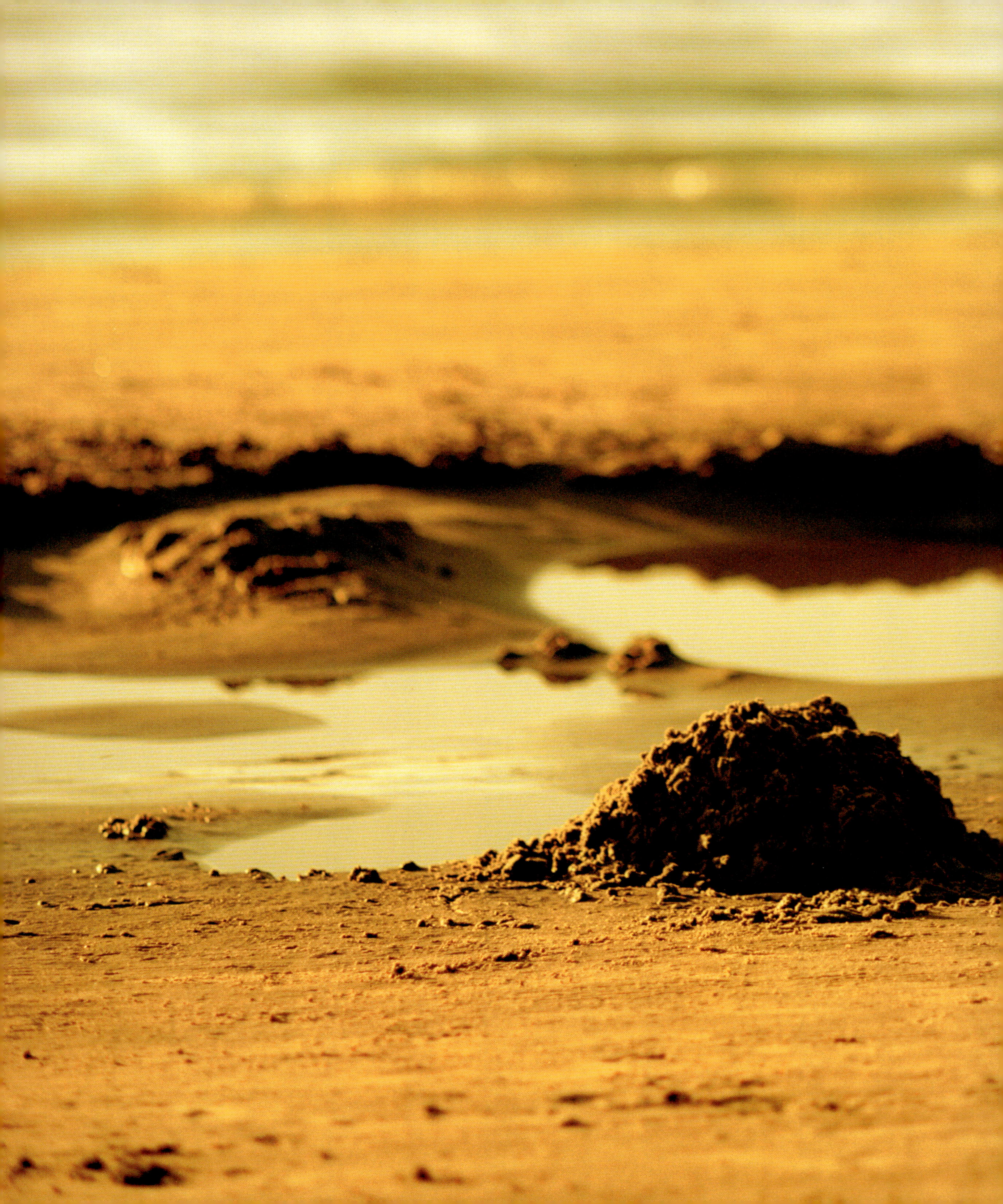

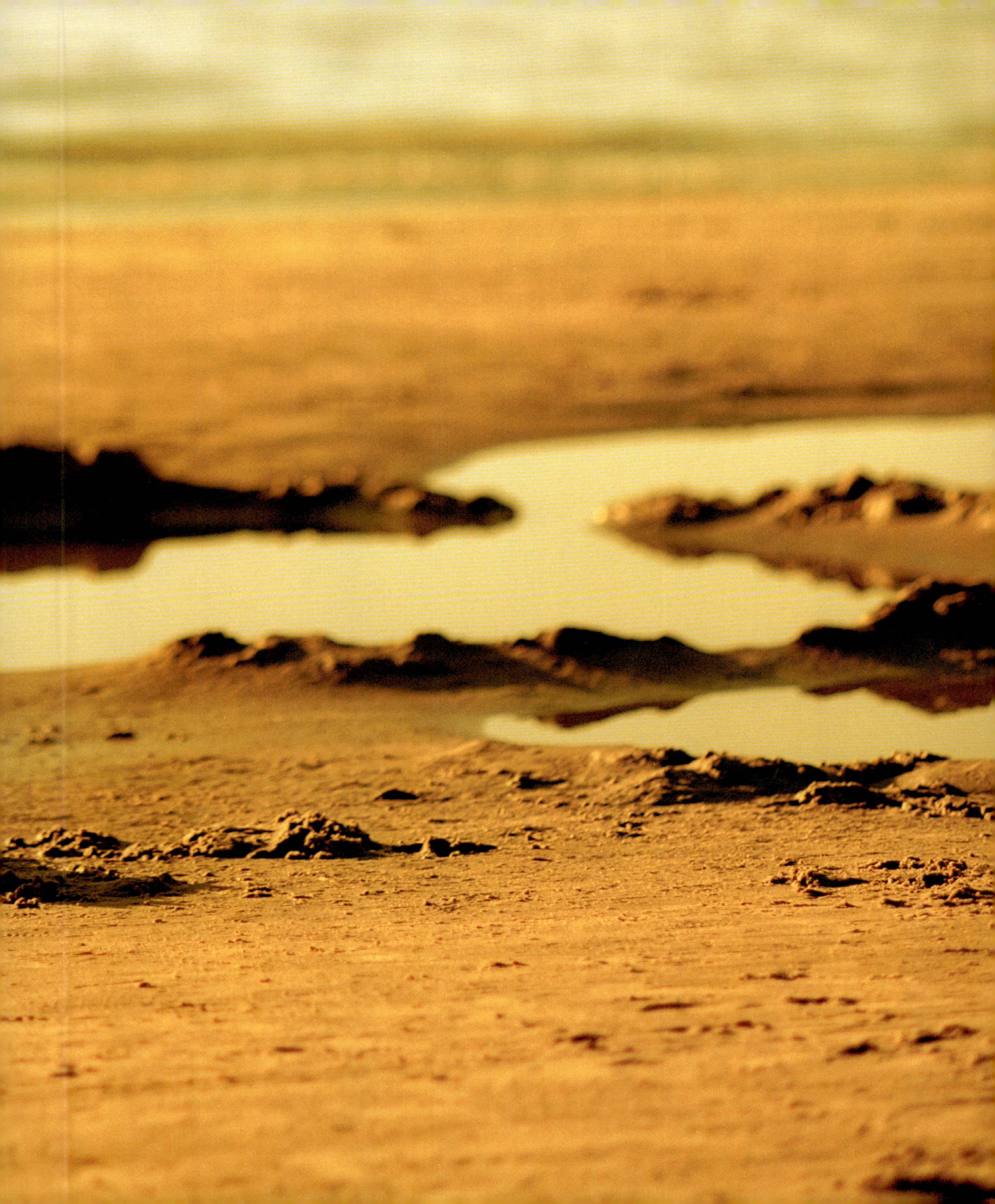

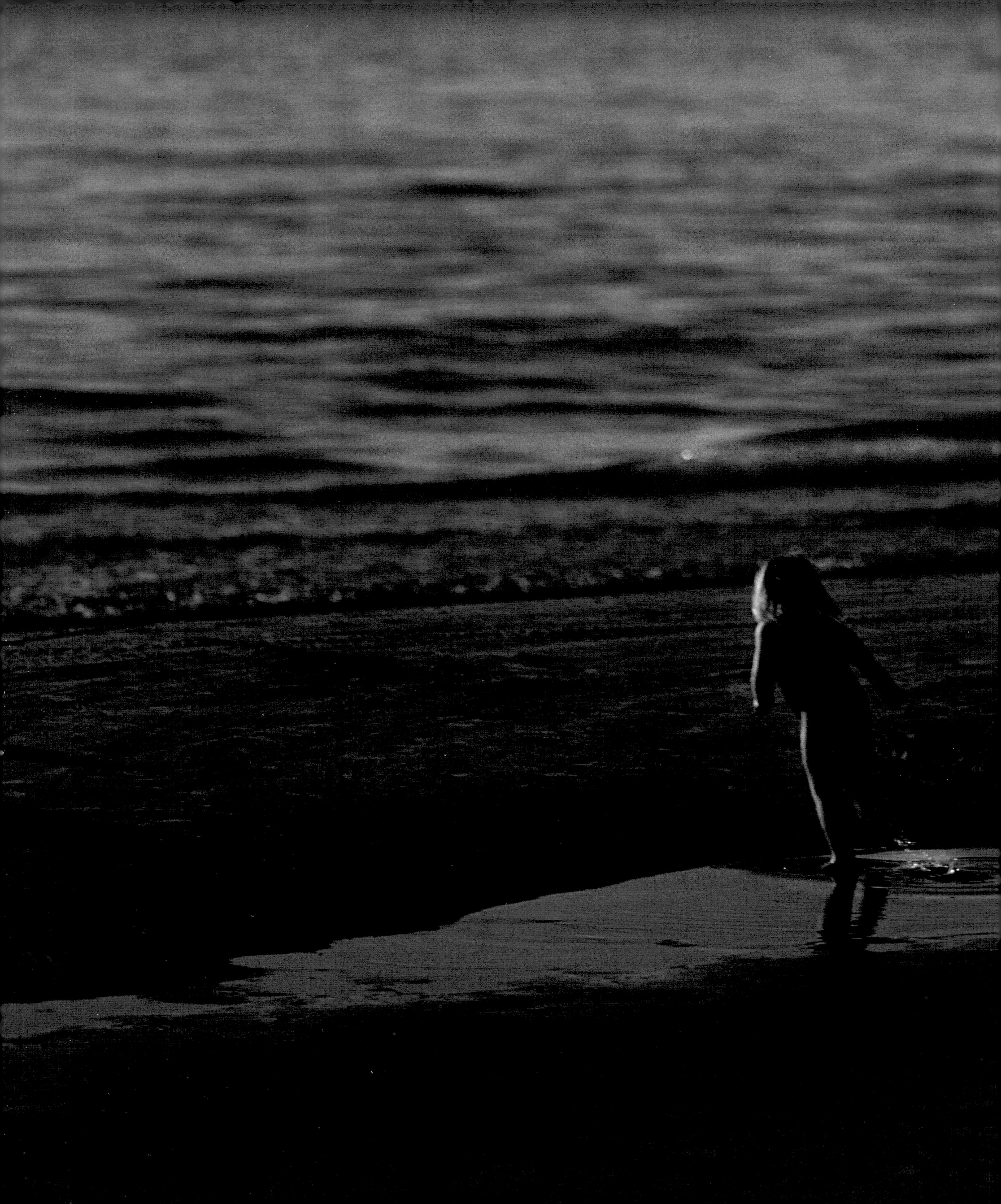

MARCO SIGNORINI

EARTHHEART

DAMIANI

Conversazione tra Marco Signorini e Angela Madesani

La cifra del lavoro di Marco Signorini non è ascrivibile a tendenze, movimenti. Il suo è un lavoro che fuoriesce dai generi canonici della fotografia: paesaggio, architettura, ritratto a maggior ragione reportage o documentazione. È interessante, al contrario, pensare, che vi sia all'interno della sua ricerca un superamento dei diversi linguaggi per giungere a una dimensione più ampia. All'interno dei suoi paesaggi sono le persone. I suoi ritratti sono immersi in contesti perlopiù naturali, in cui l'utilizzo di particolari filtri sottolinea la particolarità delle condizioni atmosferiche e geografiche, con un metodo operativo che rimanda a quello di certa pittura, pur senza trovarci di fronte a tentazioni neopittorialiste. Insomma nulla di codificato. Da un punto di vista del linguaggio, della gestione dell'immagine stessa, Signorini è immediatamente riconoscibile. Colore e bianco e nero si alternano all'interno dei diversi lavori così da creare un ritmo narrativo in cui pare di scorgere tracce di storie di persone e paesaggi. Proprio partendo di qui mi interessa chiedere spiegazioni su questa scelta dell'alternanza.

M.S. Bisogna tornare agli anni Novanta, quando con altri fotografi toscani dell'area pratese e fiorentina ci si interrogava sul senso della fotografia di paesaggio.

A.M. Quesito per molti versi ancora aperto, che in quegli anni era particolarmente attuale.

M.S. Volevamo, appunto, uscire da un discorso di genere, volevamo usare indifferentemente colore e bianco e nero, che sono entrambi astrazioni. Il mio colore, per esempio, è una sorta di monocromo, che mi piace definire cangiante: è totalmente astratto dalla realtà, anche per l'uso di filtri. Se si accompagna tutto questo con l'utilizzo del bianco e nero, che ci riporta immediatamente al passato, si dà vita a qualcosa di spiazzante. In *Echo* (1994-2005) l'utilizzo del bianco e nero non è un riferimento al passato, anzi si riferisce al contesto urbano in una dimensione temporale che potrebbe anche essere riferita al futuro.

A.M. In un'intervista[1] di qualche tempo fa, hai sottolineato il rapporto con l'arte antica, e in tal senso la fortuna di essere nato a Firenze.

M.S. Il rapporto con l'arte antica è anche l'idea di trovare delle immagini che, in qualche modo, abbiano al loro interno una costruzione che richiami proprio a quel momento artisti-co. Mi interessa cercare anche le singole immagini. Credo che produrre delle immagini che lasciano dei segni, che lasciano memoria di sé, che non siano solo una parte di un lavoro seriale sia molto importante.

A.M. A uno sguardo attento i tuoi lavori presentano numerose quanto sottili citazioni.

M.S. Cerco di lavorare parecchio per citazioni, è una forma di vizio datami dall'insegnamento, il frutto del mio lavoro nasce dallo studio, dal fatto di aver visto delle altre cose. E' un gioco con quello che si conosce, quello che si cerca di portare avanti, di portare all'attenzione, perché le cose funzionano anche o forse proprio per piccoli spostamenti.

Tornando a quanto stavo dicendo, in *Europos Centras*, il mio lavoro di esordio del 1994, ho proceduto per singole immagini. In quel particolare momento si faceva un gran parlare dei "non luoghi", un concetto coniato da Marc Augé. Molta fotografia si occupava di essi: parcheggi, aeroporti, piazzole autostradali. Ero maggiormente affascinato dai luoghi simbolici, quelli dove, tuttavia, non c'è nulla. Avevo letto che un istituto geografico aveva tracciato una sorta di diagonale attraverso l'Europa per stabilirne il centro: da Gibilterra agli Urali, da Capo Nord a Creta. Il centro cadeva nei pressi di Vilnius in Lituania, Europos Centras, appunto, dove avevano in mente di costruire un parco dell'arte[2]. Erano anni nei quali si aspirava all'unità europea. Così mi sono recato là per fare un lavoro, in un luogo dove c'era il nulla. C'erano solo bagnanti domenicali, sembravano scene dei film di Krzysztof Kieślowski o de *Il sole ingannatore* di Nikita Mikhalkov.

A.M. Parliamo del tuo rapporto con il cinema?

M.S. La mia aspirazione era quella di fare cinema. All'Accademia ho studiato Scenografia, mi interessava applicarla alla videoarte. In realtà ho studiato cinema da autodidatta. Amo molto Wim Wenders, Andrei Tarkovskij, Michelangelo Antonioni, Pier Paolo Pasolini, David Lynch. Per me il cinema è una cosa complessa, al di là della formazione di immagini. Per esempio credo che certo cinema italiano tra gli anni Sessanta e Settanta sia stato più avanti della fotografia. Per esempio nei fotogrammi dei film di Antonioni ci sono delle fotografie bellissime. Il regista ferrarese, come del resto Pasolini, aveva parlato del paesaggio attraverso le persone che si muovono in esso. Mi sento molto vicino al Pasolini di *Edipo Re*, di *Teorema*. Ci sono dei passaggi, delle soluzioni d'autore determinanti. Anche qui trovi il passaggio temporale dal bianco e nero al colore.

Il cinema, inoltre, mi interessa molto per il concetto di montaggio. Mi piace soffermarmi sui personaggi, sui ritratti, oppure allontanarmi, collocare i personaggi all'interno di un contesto. *Echo* inizia con dei particolari, poi appaiono dei ritratti, delle visioni ampie, dei campi lunghi, è un lavoro che si può dividere in diverse inquadrature usando il linguaggio cinematografico.

A.M. Roberta Valtorta, nel testo che accompagna *Echo*[3], par-

1 M. Signorini "Gli sguardi pre-esistenti" in Landscape Stories, www.landascapestories.net/interviste/marco-signorini, 09/2010.

2 Il parco è stato costruito qualche anno dopo.

3 R.Valtorta, L'eco, la casa in Marco Signorini Echo, 2006, Damiani Forward, Bologna.

la del tuo interesse nei confronti del mondo infantile.

M.S. Mi sono trovato spesso a fotografare bambini, giovani, personaggi del mondo dell'infanzia e non so se questa cosa sia casuale o se, inconsciamente, l'ho cercata. Mi piace riuscire a immaginare in un bambino le sue potenzialità, quello che sarà. Il volto degli anziani è troppo segnato dai significati, troppo carico.

A.M. In un testo[4] dedicato al tuo lavoro, la scrittrice Paola Capriolo ha parlato di geologia. Cosa ne dici?

M.S. Se c'è uno spunto geologico è per riuscire a parlare di terra, per arrivare al cuore delle cose. Anche ora sto lavorando sul tema della terra, non come luogo in cui viviamo, ma come significato della materia.

A.M. Mi pare, inoltre, che la pietra offra un collegamento al tema della memoria. Si tratta di un materiale che raccoglie una storia.

M.S. Per me è fondamentale lavorare sulla memoria, intesa non tanto come passato, come ciò che è stato, quanto come memoria primordiale, ciò che rimarrà. Mi interessa porre la mia ricerca in un orizzonte più ampio, lasciare una traccia. L'aspetto collettivo è fondamentale perché rappresenta il mio modo di intendere la fotografia. A suo tempo avevo giocato con la traduzione del termine pietra in inglese: *Stone* mi piaceva perché racchiudeva sia "sto", sto sulla pietra, qualcosa su cui si è appoggiati, sia "one" che significa l'uno, l'essere quasi primordiale. Tutti i titoli dei miei lavori sono il frutto di una scelta molto precisa. Questo libro, per esempio, ha per titolo un gioco di parole fra *Earth* e *Heart*, terra-cuore, che in inglese è un gioco enigmistico, di spostamento di consonante. Inoltre se si scrivono più volte di seguito le due parole ne scaturisce la frase *Hear the art* Senti l'arte. Questo lavoro contiene parti di *Ra, Earth* e altre opere inedite. In questo caso il cuore non è quello degli affetti, ma quello dell'intelligenza, colma di luce.

A.M. Torniamo all'aspetto sentimentale della memoria, che crea delle emozioni.

M.S. Nella mia ricerca la memoria non è riferita a luoghi particolari o a storie di persone. Il mio lavoro si rifà al reale. Fotografo quello che vedo, che si muove nel mondo, mi piace fare esperienze. Tutto il lavoro viene sottoposto a un complesso lavoro di post-produzione che riesce a rendere percipuo l'aspetto dell'immaginazione.

A.M. Sempre Paola Capriolo nel suo testo[5] parla della spiaggia di Fuerteventura come di un luogo fuori del tempo. Mi interessa parlare della dimensione temporale nella tua ricerca.

M.S. Lavoro molto su questa dimensione, che non è reale, storica, descrittiva di eventi, ma legata a un tempo assoluto. Vorrei giungere a una sorta di indefinizione.

A.M. Nei tuoi lavori percepisco un certo senso di vuoto.

M.S. In passato ho scritto delle cose sullo sguardo nel vuoto.

Quando ci si incanta a vedere qualcosa, si sovrappongono i pensieri e l'immaginazione. Si tratta di uno sguardo vuoto, ma in realtà ricco di immaginazione, che aspetta di essere colmato. Mi affascina il vuoto come vastità, come schermo. Nelle mie immagini dall'alto è come un affacciarsi nel vuoto. Ed è una cosa che provo, anch'io emotivamente, molto forte, mi dà una sensazione di paura. Il vuoto rappresenta anche la vastità di alcuni paesaggi.

A.M. La fotografia italiana non si è occupata granché di tutto questo.

M.S. L'idea dell'incanto, dell'essere nella natura, era considerato qualcosa che non si doveva fare negli anni Ottanta, era un po' troppo romantico, retorico. A dire la verità è stato uno sguardo che ha cambiato soggetto. È stato un gioco provocatorio tornare su quello che era sembrato "infotografabile". L'aspetto evocativo della natura che è molto interessante e riguarda non tanto la contemplazione della stessa, con il suo aspetto evocativo, in cui io guardo verso l'infinito e mi sento infinitamente piccolo, ma la natura in cui le persone ritrovano se stesse. Vorrei che i miei personaggi fossero ripresi non tanto come persone che contemplano, ma come persone che si guardano all'interno del paesaggio. Anche Maurice Merleau Ponty affermava: «Il mondo non è davanti a me è intorno a me». Credo che l'uomo sia partecipe di un progetto piuttosto ampio, la piccolezza non sta nelle dimensioni, quanto nell'incapacità di capire il vero senso delle cose. Quando parliamo di paesaggio, di territorio, pensiamo alla natura, pensiamo a qualcosa di primordiale che probabilmente non abbiamo mai conosciuto e non sappiamo cosa sia. Anche questo rimanda alla memoria della natura, dei paesaggi che in realtà non abbiamo mai conosciuto.

A.M. Insomma il tuo rapporto con la natura non è quello dell'islandese delle *Operette morali* di Giacomo Leopardi?

M.S. No, anche perché se osservi il mio lavoro, se c'è qualcosa di minaccioso è da parte dell'uomo nei confronti della natura e non viceversa. In tal senso si pensi al gioco di parole contenuto in *Ra*. Ra dio del sole, ma Ra simbolo del radio con un chiaro richiamo alla radioattività. "Ra" è anche il diminutivo di radial, i pneumatici, di cui nel lavoro c'è traccia. Qui è tutto giocato sulla luce, su questo aspetto dell'umano pilotato dalle macchine minacciose con i fari accesi, come in un'immagine cinematografica alla David Lynch. La natura non è incontaminata, è come nella realtà. Quel lavoro è stato realizzato tutto in meno di cinquanta minuti, è costituito da pochissime immagini.

A.M. *Ra* quando nasce?

M.S. É nato, nel 2008. Sono andato alle isole Frisone, tra il nord della Germania e la Danimarca, per cercare dei luoghi che già avevo in mente. Mi interessava la luce particolare che avrei potuto trovarvi. Sono territori in continua trasformazione a causa della bassa marea. Volevo fare un lavoro sulla contemporaneità. In quelle spiagge potevo fotografare le auto che passavano con i fari accesi. Ho ruotato così attorno a temi

4 P.Capriolo, 2009 in Fotografia europea. Eternità. Il tempo dell'immagine. (Catalogo della mostra Reggio Emilia, 30 aprile-7giugno 2009) a cura di E. Grazioli, 2009, Mondadori Electa, Milano, pag. 150-151.
5 P.Capriolo, op.cit., 2009.

forti: la luce, la vastità del paesaggio, attraverso dei personaggi chiave, la madre e i figli, che si muovono nella natura. La madre, la terra, il sole, la luce, i figli, la famiglia è tutto un concatenamento di elementi.

A.M. In *Earth* (2006), invece, la presenza umana ricopre un ruolo più marginale.

M.S. È un lavoro su luoghi diversi, realizzato in tempi diversi, mentre in *Ra* c'è la rotazione dello sguardo, tutto ciò che è accaduto in pochissimi metri, nello stesso luogo e nello stesso tempo. *Earth* è composto da immagini che ho realizzato nelle due isole di Fuerteventura e Lanzarote. In certi punti pare di trovarsi di fronte a un paesaggio lunare. Andando a lavorare in un posto, a volte si trovano degli elementi che diventano la sostanza del lavoro. Non volendo fare un lavoro descrittivo, non volendo documentare dei luoghi, mi affido a quello che trovo, per poter strutturare il racconto in chiave cinematografica. Ho trovato delle pietre che a mio parere potevano rappresentare una sorta di caos primordiale. Mi pareva una dimensione vicina a quella della Land Art.

A.M. La tua ricerca, in cui l'uomo è coprotagonista ha una valenza sociale?

M.S. La mia idea è quella di pensare che, attraverso una certa bellezza, una certa idea della natura, un certo tipo di immagini e colori, in realtà ci sia una riflessione sulla propria esistenza e sul proprio modo di stare al mondo per tornare alle domande di sempre: chi sono, che ci faccio qui? L'elemento di contemporaneità è sempre emotivo. Questo non è il paesaggio contemporaneo alla "New Topographics" americana, dove c'è il segno del fuoco lasciato, è qualcosa che, piuttosto, dovrebbe coinvolgere. Mi pare che il paesaggio, la natura, intesi come pianeta siano un elemento fortemente sociale. La natura dovrebbe divenire un elemento di precipuo interesse per la politica. Dietro il tramonto, dietro la bellezza di paesaggio onirico, in realtà ci sono anche delle minacce.

A.M. Walter Guadagnini[6] ha scritto che la tua ricerca si fonda sul rapporto ambiguo tra ciò che è visibile e ciò che è inesprimibile, tra il visto e il non detto.

M.S. Nella fotografia mi interessa molto quello che è apparentemente non visibile. Cioè il visibile che diventa il tramite dell'invisibile. L'immagine, il soggetto diventano un tramite per parlare di qualcos'altro che non è il soggetto ma è quello che ci sta dietro. La fotografia potrebbe avere una valenza enigmistica. È come una sorta di anagramma: si fotografano le cose, si riproducono e la realtà si ricompone apparentemente nello stesso posto sulla pellicola. Come una rivelazione. Da questo punto di vista mi interessa anche la forma del rebus, dove si ha un'immagine, che per risolvere bisogna descrivere, ma il rebus diventa una risultante che non ha nessuna attinenza con l'immagine dal punto di vista del significato della frase. È il passaggio fra il vero e la realtà, quella che Thomas Ruff chiama la realtà reale, di primo grado. Con la fotografia io faccio riferimento a una realtà di primo grado appunto. Fo-

tografo muovendomi, viaggiando e camminando. Sono uno *street photographer*, non produco immagini al computer, c'è sempre una partenza dalla realtà.

A.M. Davanti alle tue opere si può parlare di visionarietà.

M.S. Penso, piuttosto, che si possa parlare di visione. In un mio testo che accompagnava la mostra *Idea di Metropoli*[7], scrivevo che l'uomo fotografo è un passante da sempre atteso da sguardi pre-esistenti. Non metto in moto la visione perché guardo, fotografo.

La visione c'è, esiste, il sogno è la dimostrazione che non è soltanto attraverso lo sguardo, la vista che si hanno delle visioni, è un continuo rimando fra menti altre. La visione è un continuum tra il mio sguardo e ciò che ti guarda. Il mio è un tentativo di mostrare una visione che già c'è. Mi piace cercare fra le cose più banali che si rifanno alla realtà, cercare un significato che porti a qualcosa che vada oltre.

Il paesaggio in *Earth*, ad esempio, è molto lunare e si percepisce una dimensione onirica. Mi piace il paesaggio meno strutturato, più desolato. E' un orizzonte che a noi, in Italia, manca. Il mio è un tentativo di creare un effetto che ti faccia riflettere sul tempo, che ti possa sembrare altro rispetto a quello che sto mostrando.

A.M. Possiamo parlare della valenza che hanno per te i libri, del loro aspetto progettuale.

M.S. I miei libri non sono semplicemente dei cataloghi dei lavori, quanto piuttosto dei lavori autonomi. L'idea del libro è un punto fondamentale, è l'opera, l'opera cinematografica, è il film che ti permette di trasmettere il significato che, all'interno di un singolo lavoro, difficilmente riesce a venir fuori. Mi piace lavorare per episodi, che dovrebbero essere l'essenza dei vari progetti. Mi piacciono molto i libri strutturati per racconti, mi piace che vari racconti, messi insieme riescano a trovare un significato ulteriore, a divenire una storia più strutturata. E' come dispensare lentamente le cose, lasciando decantare i lavori, magari per anni, per poi utilizzarli in una nuova chiave come sta capitando per questa mostra. Quest'operazione offre la dimensione reale del mutamento. Dà la possibilità di rileggere quello che si è fatto nel passato e, in qualche modo, di riattualizzarlo.

---

6 W.Guadagnini, Enigmi e incanti (catalogo della mostra presso la galleria Metronom di Modena, 4 aprile al 6 giugno 2009).

7 La mostra ha avuto luogo nel 2002 al Museo di Fotografia Contemporanea Villa Ghirlanda, Cinisello Balsamo (Milano).

Conversation entre Marco Signorini et Angela Madesani

Le langage des œuvres de Marco Signorini ne s'insère dans aucune tendance ni aucun mouvement. Sa recherche se situe hors des genres canoniques de la photographie: le paysage, l'architecture, le portrait ou à plus forte raison le reportage ou la photo documentaire. Au contraire, il est intéressant de penser qu'il existe à l'intérieur de sa recherche un dépassement des différents langages pour atteindre une dimension plus vaste. Ses paysages contiennent des personnes. Ses portraits sont immergés dans des environnements le plus souvent naturels, dans lesquels l'utilisation de filtres spécifiques souligne la particularité des conditions atmosphériques et géographiques, selon un mode opératoire qui renvoie à un certain type de peinture, sans que l'on soit pour autant confrontés à des tentations néo-pictorialistes. En bref, rien de codifié. Du point de vue du langage, de la gestion de l'image en tant que telle, Signorini est immédiatement reconnaissable. La couleur et le noir et blanc s'alternent à l'intérieur de ses différentes œuvres de façon à créer un rythme narratif dans lequel il est possible de percevoir des ébauches d'histoires de personnes et de paysages. J'aimerais prendre cela comme point de départ pour notre conversation, ce choix de l'alternance.

M.S. Pour cela il faudrait remonter aux années '90, à l'époque où, avec d'autres photographes de la région de Prato et de Florence, nous nous interrogions sur le sens de la photo de paysage.

A.M. Une question qui sous plus d'un aspect demeure ouverte, et qui était particulièrement actuelle dans ces années.

M.S. Nous voulions précisément sortir du débat sur les genres et voulions utiliser indifféremment la couleur et le noir et blanc, qui sont tous deux des abstractions. Ma couleur, par exemple, est une sorte de monochrome que j'aime définir comme chatoyant: elle est totalement étrangère à la réalité, grâce à l'utilisation de filtres. Si l'on marie à cela l'utilisation du noir et blanc, qui nous renvoie immédiatement au passé, on donne vie à quelque chose de déconcertant. Dans *Echo* (1994-2005), l'utilisation du noir et blanc ne renvoie pas au passé mais plutôt à l'environnement urbain dans une dimension temporelle qui pourrait aussi renvoyer au futur.

A.M. Il y a quelque temps, tu as souligné dans une interview[1] ton rapport avec l'art antique, et, dans ce sens, la chance que tu as eue d'être né à Florence.

M.S. Mon rapport avec l'art antique implique aussi l'idée de trouver des images qui, d'une certaine façon, contiennent une construction renvoyant à un moment artistique précis. Je suis aussi intéressé par la recherche d'images individuelles. Je crois qu'il est très important de produire des images qui laissent des signes, qui marquent la mémoire, et qui ne soient pas seulement une partie d'un travail sériel.

A.M. A un regard attentif tes travaux révèlent des citations aussi nombreuses que subtiles.

M.S. Je travaille beaucoup avec les citations, c'est une sorte de manie qui me vient de l'enseignement. Le fruit de mon travail naît de l'étude, du fait d'avoir vu d'autres choses. C'est un jeu avec ce que l'on connaît, ce que l'on s'efforce de poursuivre, de soumettre à l'attention, car les choses fonctionnement aussi ou peut-être seulement par petits déplacements. Pour revenir à ce que je disais, dans *Europos Centras*, mon premier travail daté de 1994, j'ai procédé par images individuelles. A l'époque on parlait beaucoup des "non-lieux", un concept forgé par Marc Augé. Beaucoup de photographes s'en occupaient: parkings, aéroports, aires d'autoroutes. J'étais plus fasciné par les lieux symboliques, ceux où, malgré tout, il n'y a rien. J'avais lu qu'un institut géographique avait tracé une sorte de diagonale à travers l'Europe pour en déterminer le centre: de Gibraltar à l'Oural, du Cap Nord à l'île de Crète. Le centre se trouvait près de Vilnius, en Lituanie: Europos Centras, donc, où l'on projetait de construire un parc consacré à l'art[2]. C'était les années où l'on aspirait à l'unité européenne. Ainsi je me suis rendu là-bas pour faire un travail, à un endroit où il n'y avait rien. On n'y croisait que des baigneurs du dimanche, dans des scènes qui rappelaient les films de Krzysztof Kieślowski ou *Soleil trompeur* de Nikita Mikhalkov.

A.M. Quelques mots sur tes rapports avec le cinéma?

M.S. J'aspirais à faire du cinéma. A l'Académie j'ai fait des études de scénographie que j'aimais conjuguer à l'art vidéo. En réalité, j'ai étudié le cinéma en autodidacte. J'aime beaucoup Wim Wenders, Andreï Tarkovski, Michelangelo Antonioni, Pier Paolo Pasolini, David Lynch. Je crois que le cinéma est une chose complexe, au-delà de la question de la création des images. Par exemple, je crois qu'un certain cinéma italien des années '60 et '70 a été plus loin que la photographie. Les photogrammes d'Antonioni contiennent de très belles photos. Le réalisateur de Ferrare, du reste comme Pasolini, parlait du paysage à travers les personnes qui y évoluent. Je me sens très proche du Pasolini d'*Œdipe-Roi* et de *Théorème*. Ces films contiennent des passages, des solutions d'auteur déterminantes. Nous y retrouvons aussi le passage temporel du noir et blanc à la couleur. Par ailleurs, le cinéma m'intéresse beaucoup en vertu du concept de montage. J'aime m'arrêter sur les personnages, sur les portraits, ou bien m'éloigner, situer les personnages à l'intérieur d'un contexte. *Echo* commence avec des détails, viennent ensuite

---

1 M. Signorini "Gli sguardi pre-esistenti" ["Les regards préexistants"] in Landscape Stories, www.landascapestories.net/interviste/marco-signorini, 09/2010.

2 Le parc a été construit quelques années plus tard.

des portraits, de vastes visions, des champs profonds: ce travail peut se décomposer en plusieurs cadrages en suivant le langage cinématographique.

A.M. Roberta Valtorta, dans le texte qui accompagne *Echo*[3], parle de ton intérêt pour le monde de l'enfance.

M.S. Il m'est souvent arrivé de photographier des enfants, des jeunes, des personnages du monde de l'enfance. Je ne sais pas si cela s'est produit par hasard ou bien si, inconsciemment, je l'ai cherché. Dans un enfant, j'aime réussir à imaginer ses potentialités, ce qu'il deviendra. Le visage des personnes âgées est trop lourd de significations, trop chargé.

A.M. Dans un texte[4] consacré à ton œuvre, l'écrivain Paola Capriolo a évoqué la géologie. Qu'en penses-tu?

M.S. S'il existe un rapport avec la géologie, c'est pour parler de la terre, pour atteindre le cœur des choses. Je travaille encore actuellement sur le thème de la terre, non pas en tant que lieu dans lequel nous vivons, mais comme signifié de la matière.

A.M. Par ailleurs, il me semble que la pierre nous offre un lien avec le thème de la mémoire. Il s'agit d'un matériau qui contient une histoire.

M.S. Pour moi, il est fondamental de travailler sur la mémoire non seulement dans le sens du passé, de ce qui a été, mais surtout dans celui de mémoire primordiale, de ce qui restera. Ce qui m'intéresse, c'est situer ma recherche dans un horizon plus vaste, laisser une trace. L'aspect collectif est fondamental car il représente ma façon de comprendre la photographie. Dans le passé, j'ai joué avec la traduction du mot qui signifie "pierre" en anglais, "stone": ce terme me plaisait car il contenait à la fois "sto", comme dans "sto sulla pietra", et "one", qui signifie "un", l'être primordial. Tous les titres de mes travaux sont le fruit d'un choix très précis. Le titre du présent livre, par exemple, se base sur un jeu de mot entre "earth" et "heart", terre-cœur, en anglais, basé sur un déplacement de consonne. Ce travail contient des parties de *Ra*, de *Earth* et d'autres œuvres inédites. Dans ce cas précis, le cœur n'est pas celui des affects, mais celui de l'intelligence, qui est pleine de lumière.

A.M. Revenons à l'aspect sentimental de la mémoire, qui crée des émotions.

M.S. Dans ma recherche, la mémoire ne se réfère pas à des lieux particuliers ou à des histoires individuelles. Mon travail part de la réalité. Je photographie ce que je vois, ce qui bouge dans le monde. J'aime faire des expériences. L'ensemble du travail est soumis à un processus de post-production complexe qui permet de rendre perceptible l'aspect imaginatif.

A.M. Dans le même texte[5], Paola Capriolo parle de la plage de Fuerteventura comme d'un lieu hors du temps. J'aimerais parler de la dimension temporelle de ta recherche.

M.S. Je travaille beaucoup sur cette dimension, non pas sur la dimension réelle, historique, où les événements ont lieu, mais en rapport avec un temps absolu. J'aimerais arriver à une sorte d'"indéfinition".

A.M. Dans tes œuvres, je perçois une certaine sensation de vide.

M.S. Dans le passé j'ai écrit des choses sur le regard dans le vide. Lorsqu'on s'oublie en regardant quelque chose, les pensées et l'imagination se superposent. Il s'agit d'un regard vide qui, cependant, est riche d'imagination: un vide qui attend d'être comblé. Le vide me fascine en tant qu'immensité, en tant qu'écran. Dans mes images prises de haut, on a comme une avancée dans le vide. C'est une émotion que j'éprouve moi-même de façon très vive, le vide me procure une sensation de peur. Le vide représente aussi l'immensité de certains paysages.

A.M. La photographie italienne ne s'est pas beaucoup occupée de tout cela.

M.S. Dans les années '80, l'idée de la fascination, de se trouver dans la nature, était considéré comme une chose à éviter – c'était un peu trop romantique et rhétorique. En réalité il s'agit d'un regard qui a changé de sujet. Renouer avec ce qui semblait "inphotographiable" était un jeu provocateur. L'aspect évocateur de la nature est très intéressant et ne concerne pas tant la contemplation de la nature, avec son côté suggestif, où je regarde l'infini en me sentant infiniment petit, que la nature dans laquelle les individus se retrouvent eux-mêmes. Je voudrais que mes personnages soient perçus non pas comme des personnes qui contemplent, mais comme des individus qui se regardent à l'intérieur du paysage. Merleau-Ponty affirmait: "Le monde n'est pas devant moi mais autour de moi." Je crois que l'homme participe d'un projet plutôt vaste et que la petitesse ne réside pas dans les dimensions mais dans l'incapacité de comprendre le véritable sens des choses. Quand nous parlons de paysage, de territoire, nous pensons au monde naturel, à quelque chose de primordial que nous n'avons probablement jamais connu et dont nous ignorons la nature. Cela aussi renvoie à la mémoire du monde naturel et des paysages que nous n'avons jamais connus.

A.M. Ton rapport avec la nature est-il comparable à celui de l'Islandais des *Petites œuvres morales* de Giacomo Leopardi?

M.S. Non. Si l'on observe mes photos, les seules menaces présentes sont celles que l'homme fait peser sur la nature, et non l'inverse. Dans ce sens, pensons au jeu de mots que contient *Ra*. Râ, dieu du soleil, mais aussi Ra en tant que symbole du radium, avec une claire allusion à la radioactivité. "Ra" est aussi le diminutif de "radial", en référence au pneumatiques, qui sont évoqués dans les photos. Dans ce travail, tout se joue sur la lumière, sur cet aspect de l'humain piloté par des machines menaçantes ayant leurs phares allumés, comme dans une image cinématographique à la David Lynch. La nature n'est pas indemne de contamination, elle est comme dans la réalité. Ce travail a été entièrement réalisé en moins de cinquante minutes, il se compose de très peu d'images.

3 R.Valtorta, "L'eco, la casa" in Marco Signorini, Echo, 2006, Damiani Forward, Bologna.
4 P.Capriolo, 2009 in Fotografia europea. Eternità. Il tempo dell'immagine.
(Catalogue de l'exposition de Reggio Emilia, 30 avril-7 juin 2009) sous la direction de
E. Grazioli, 2009, Mondadori Electa, Milano, pp. 150-151.
5 P.Capriolo, op.cit., 2009.

A.M. De quand date *Ra*?

M.S. Ce travail a vu le jour en 2008. Je me suis rendu dans les îles de la Frise, entre le nord de l'Allemagne et le Danemark, à la recherche de lieux que j'avais déjà à l'esprit. J'étais intéressé par la lumière particulière que j'aurais pu y trouver. Ce sont des territoires en constante mutation à cause de la marée basse. Je voulais faire un travail sur la contemporanéité. Depuis ces plages, je pouvais photographier les voitures qui passaient avec leurs phares allumés. Ainsi, j'ai gravité autour de thèmes forts: la lumière, la vastitude du paysage, à travers des personnages clés, la mère et ses enfants, qui évoluent dans la nature. La mère, la terre, le soleil, la lumière, les fils, la famille: c'est un enchaînement d'éléments.

A.M. Dans *Earth* (2006), en revanche, la présence humaine a un rôle plus marginal.

M.S. C'est un travail sur des lieux différents, réalisé avec des délais différents. *Ra* met en scène une rotation du regard et tout ce qui est arrivé en quelques mètres, au même moment et au même endroit. *Earth* se compose d'images que j'ai capturées sur les deux îles de Fuerteventura et Lanzarote. A certains endroits, on a l'impression d'être face à un paysage lunaire. Parfois, lorsque l'on se rend à un endroit pour travailler, on tombe sur des éléments qui deviennent la substance du projet. Ne voulant pas faire un travail descriptif ou documenter des lieux, je fais confiance à ce que je trouve pour pouvoir structurer la narration sur un mode cinématographique. J'ai trouvé des pierres qui selon moi pouvaient représenter une sorte de chaos primordial. Je trouvais que cette dimension était proche du Land Art.

A.M. Ta recherche, dans laquelle l'homme est co-protagoniste, a-t-elle une valence sociale?

M.S. Je crois qu'à travers une certaine beauté, une certaine idée de la nature, un certain type d'image et de couleurs, il est possible d'effectuer une réflexion sur sa propre existence et sur sa propre manière d'être au monde pour revenir aux questions de toujours: qui suis-je, que fais-je ici? L'élément de contemporanéité est toujours de nature émotive. Ce n'est pas le paysage contemporain à la "New Topographics" américaine, où il demeure toujours une trace de la présence humaine, mais plutôt quelque chose qui devrait impliquer l'observateur. Je crois que le paysage et la nature, compris comme faisant partie de la planète, sont des éléments à forte valence sociale. La nature devrait devenir un objet d'intérêt central pour la politique. En réalité, le coucher de soleil et la beauté onirique du paysage cachent aussi des menaces.

A.M. Walter Guadagnini[6] a écrit que ta recherche se fonde sur le rapport ambigu entre ce qui est visible et ce qui est inexprimable, entre le vu et le non-dit.

M.S. En photographie je suis extrêmement intéressé par ce qui est apparemment invisible. C'est-à-dire par le visible qui devient l'intermédiaire de l'invisible. L'image et son objet deviennent un moyen de parler d'autre chose, d'une chose qui n'est pas l'objet mais ce qui se trouve derrière lui. La photographie pourrait avoir une valence énigmatique. C'est comme une sorte d'anagramme: on photographie les choses, on les reproduit, et la réalité se recompose apparemment au même endroit sur la pellicule. C'est comme une révélation. Je suis également intéressé par la forme du rébus, où l'image demande à être déchiffrée: le sens qui en résulte n'a aucun rapport avec l'image en soi. C'est le passage entre le vrai et la réalité, que Thomas Ruff appelle la réalité réelle, du premier degré. Mes photos se réfèrent à une réalité du premier degré. Je photographie en me déplaçant, en voyageant et en marchant. Je suis un "street photographer": je ne réalise pas d'images par ordinateur, je pars toujours de la réalité.

A.M. Tes œuvres ont un caractère visionnaire.

M.S. Je parlerais plutôt de visions. Dans un texte que j'ai écrit pour l'exposition *Idea di Metropoli*[7], j'ai écrit que l'homme photographe est un passant attendu depuis toujours par des regards préexistants. Je ne déclenche pas la vision parce que je regarde ou photographie. La vision est là, elle existe. Le rêve fournit la preuve que ce n'est pas seulement à travers le regard, la vue, que l'on a des visions, mais par un renvoi continu entre différents esprits. La vision est un continuum entre mon regard et ce qui me regarde. Je m'efforce de montrer une vision qui est déjà là. J'aime fouiller parmi les choses les plus banales qui rappellent la réalité, chercher un sens qui nous mène à un dépassement. Dans *Earth*, par exemple, le paysage est lunaire et laisse transparaître une dimension onirique. J'aime les paysages moins structurés, plus désolés. C'est un horizon qui nous manque, en Italie. Je m'efforce de créer un effet qui suscite une réflexion sur le temps, qui puisse paraître autre par rapport à ce que je montre.

A.M. Quelques mots au sujet de la valeur qu'ont les livres pour toi, de leur aspect projectuel?

M.S. Mes livres ne sont pas simplement des catalogues d'images, mais un projet à part entière. L'idée du livre est un élément fondamental: c'est l'œuvre, l'œuvre cinématographique, le film qui te permet de transmettre le sens qui, à l'intérieur d'une seule image, ne transparaît que difficilement. J'aime travailler par épisodes qui devraient être l'essence des différents projets. J'aime beaucoup les livres structurés de façon narrative, et j'aime que les différentes narrations réunies acquièrent une signification supplémentaire en formant une narration plus structurée. Cela est comparable à une lente décantation les images, qui peut durer des années, pour ensuite les utiliser selon un angle nouveau, comme c'est le cas pour le présent projet. Cette opération nous donne la dimension réelle du changement. Elle permet de relire ce que l'on a fait dans le passé et, d'une certaine façon, de le réactualiser.

---

6 W.Guadagnini, Enigmi e incanti (catalogue de l'exposition à la galerie Metronom de Modène, du 4 avril au 6 juin 2009).

7 L'exposition a eu lieu en 2002 au Musée de Photographie Contemporaine, Villa Ghirlanda, Cinisello Balsamo (Milan).

Conversation between Marco Signorini and Angela Madesani

The key stylistic feature of Marco Signorini's work cannot be attributed to trends or movements. His work escapes the generic canons of photography: landscape, architecture, portrait or even reportage or documentation. It is interesting, on the other hand, to think that within his work there is an overtaking of various forms of expression in an effort to reach a broader dimension. There are people in his landscapes. His portraits are immersed in mostly natural places in which the use of special filters underlines the uniqueness of the atmospheric and geographical conditions, employing a technique that recalls that of certain paintings, though without succumbing to neo-pictorial temptations. In short, there is nothing coded. From the point of view of the language, of the handling of the image itself, Signorini is instantly recognisable. Colour and black and white alternate within various works in order to create a narrative pattern where one seems to glimpse traces of stories of people and landscapes. Taking precisely this as a point of departure I am keen to hear an explanation regarding this decision to alternate.

M.S. We must go back to the Nineties when I, and other Tuscan photographers from the area around Prato and Florence, questioned ourselves about the meaning of landscape photography.

A.M. This question, which was especially relevant in those years is, in many ways, still open-ended.

M.S. We wanted to depart from a discourse of genre. We wanted to use colour and black and white, which are both abstractions, equally. My colour, for instance, is a kind of monochrome which I like to define as iridescent. It is totally abstracted from reality, partly due to the use of filters. If all of this is associated with the use of black and white, which catapults us straight back to the past, something is generated which catches you unprepared. In *Echo* (1994-2005) the use of black and white is not a reference to the past. Indeed, it refers to the urban context in a temporal dimension which might also allude to the future.

A.M. In an interview[1] a while back you emphasised the relationship with ancient art and how lucky you were to be born in Florence in this sense.

M.S. The relationship with ancient art also implies coming up with images which somehow contain a structure which re-calls that artistic period. I am interested in seeking single images too. I believe that it is very important to produce images which leave traces, which leave a memory of themselves and which are not just one in a series of works.

A.M. If one looks closely one sees that your works contain numerous, if subtle, quotes.

M.S. I try to work by quotes a lot. It is a kind of habit I picked up from teaching. The fruit of my work comes from studying, from the fact of having seen other things. It is a game consisting of what one knows, what one tries to foster, to bring to the fore, because things also, and perhaps only, work by means of subtle displacements.

Going back to what I was saying before, in *Europos Centras,* my debut piece of 1994, I worked on single images. At that particular time there was a great deal of talk about "non places", a concept coined by Marc Augé. A lot of photography dealt with these non-places: car parks, airports, motorway service stations. I was more fascinated by symbolic places, those places which, nevertheless, contain nothing. I had read that a geographical institute had traced a sort of diagonal line across Europe in order to establish its centre: from Gibraltar to the Urals, from the North Cape to Crete. The centre fell near Vilnius in Lithuania, Europos Centras, where they were thinking of building an art park[2]. In those years when there was still the dream of European unity. So I went there to do some work, in a place which contained nothing. There were only Sunday bathers. It looked like scenes from the film of Krzysztof Kieślowski or *Burnt by the Sun* by Nikita Mikhalkov.

A.M. Shall we talk about your relationship with the cinema?

M.S. My ambition was to be a film-maker. When I was at the Academy I studied set design and I was interested in applying it to video art. Actually I am a self-taught cineaste. I love Wim Wenders, Andrei Tarkovskij, Michelangelo Antonioni, Pier Paolo Pasolini, David Lynch. For me cinema was a complex thing, something more than the production of images. For example, I believe that a certain genre of Italian cinema in the Sixties and Seventies was ahead of photography. For instance, the stills of Antonioni's films contain beautiful photographs. The director, a native of Ferrara, like Pasolini after all, spoke of the landscape through the people who moved through it. I feel a close affinity to the Pasolini of Oedipus Rex, of Teorema. They contain landscapes, artist's choices, which are key. There too you find a temporal landscape ranging from black and white to colour.

Moreover, I am very interested in cinema due to the concept of cutting. I like to linger on the characters, on the portraits, or else stand back and place the characters within a context. *Echo* begins with details, which are followed by portraits, broad visions, long fields. It is a work which can be divided into various frames using cinematic language.

A.M. In the text which accompanies *Echo*[3] Roberta Valtorta talks about your interest in the world of childhood.

1 M. Signorini "Gli sguardi pre-esistenti" in Landscape Stories, www.landscapestories.net/interviste/marco-signorini, 09/2010.

2 The park was built a few years later.
3 R.Valtorta, L'eco, la casa in Marco Signorini Echo, 2006, Damiani Forward, Bologna.

M.S. I have often happened to photograph children, young people and characters from the world of children, and I don't know if this thing is random or if I have pursued it unconsciously. I like to succeed in imagining a child's potential, what it will become. The faces of old people are too scored with meanings, too loaded.

A.M. In a text[4] dedicated to your work writer Paola Capriolo spoke of geology. What can you say about this?

M.S. If there is a geological element it is there so that I can talk about the earth, in order to reach the core of things. Now I am working on the theme of the earth, not as a place we live in, but as the meaning of matter.

A.M. Moreover, it seems to me that stone provides a link with the theme of memory. It is a material that contains a story.

M.S. For me it is essential to work on memory, and by this I mean not so much the past or what once was but primordial memory or what will remain. I am interested in positioning my work in a wider context and leaving a trace. The collective aspect is essential because it represents my way of seeing photography. Originally I played with the translation of the term stone: I liked *Stone* because it contained both the Italian "sto" or "I am", I am on the stone, something on which you lean, and "one", the near primordial state. All the titles of my work are the result of a painstaking choice. The title of this text, for instance, is a play on words that involves *Earth and Heart*, which is a puzzle created by shifting the consonants. Moreover, if you write the words over and over again it produces the sentence "Hear the Art". This work contains part of *Ra, Earth* and other previously unpublished works. In this case the heart is not the heart of affections but that of intelligence, brimming with light.

A.M. Let's return to the sentimental aspect of memory, which generates emotions.

M.S. In my work memory does not refer to particular places or to people's stories. My work is related to what is real. I photograph what I see, what moves in the world. I like having experiences. All of my work undergoes a complex post-production process which succeeds in rendering the aspect of the imagination extraordinary.

A.M. Once again Paola Capriolo talks[5] about the beach of Fuerteventura as timeless place. I am interested in talking about the temporal dimension in your work.

M.S. I work a great deal on this dimension, which is not real, historical or descriptive of events, but linked to an absolute time. I aim to achieve a sort of "indefinition".

A.M. In your works I perceive a certain sense of emptiness.

M.S. In the past I wrote stuff about gazing into space. When you stop and stare at something thoughts and the imagination are superimposed. You are gazing into space but, in reality, the imagination runs riot and tries to fill the void. I am fascinated by emptiness as an expanse, as a screen. In my pictures taken from above there is the sense that you are looking out into a void. It is something I feel emotionally and very powerfully and it fills me with a sense of fear. The void also represents the vastness of some landscapes.

A.M. Italian photography hasn't engaged much with any of this.

M.S. The idea of stopping and staring, of being in nature, was regarded as something that shouldn't be done in the Eighties. It was a little too romantic and rhetorical. To tell the truth it was a way of looking that changed the subject. Returning to what had seemed "unphotographable" was a provocation. The evocative aspect of nature, which is extremely interesting and concerns not so much the contemplation of the same, with its evocative aspect, where I look towards infinity and feel infinitely small, but nature where people can feel in their element. I want my characters to be photographed not so much as people in the act of contemplation but as people who are looking at themselves in a landscape. Maurice Merleau Ponty also said: "The world is not in front of me. It is around me". I believe that man is a participant in quite a large project. Smallness is not a question of size but the inability to understand the true meaning of things. When we talk about landscape and about territory, we think about nature, we think about something primordial that we have probably never known and that we haven't the least idea about. This too is connected to the memory of nature and of landscapes which we have actually never known.

A.M. In short, your relationship with nature is not the Icelandic one described in Giacomo Leopardi's *Moral Tales*?

M.S. No, also because if you observe my work, you will see that if there is a threatening element it originates in men and is directed towards nature and not vice versa. In this sense think of the play on words in Ra. Ra = sun god, but Ra is also the chemical symbol of radium with a clear reference to radioactivity. "Ra" is also short for radials, tyres, of which there is a trace in my work. Here it all hinges on light, on that aspect of man driven by ominous cars with blazing headlights, like in a David Lynch movie image. Nature is not unspoilt, it is how it is in real life. That work was produced in under fifty minutes. It consists of just a handful of pictures.

A.M. When was *Ra* done?

M.S. It was done in 2008. I went to the Frisian islands, between the north of Germany and Denmark, in order to look for the places I already had in my head. I was interested in the special quality of light I could find there. These territories are in perpetual transformation due to the low tide. I wanted to produce a work on contemporaneity. On those beaches I was able to photograph the cars passing with lit headlights. Consequently, I rotated around powerful themes: light, the vastness of the landscape, by means of key characters, the mother and the children, who move through nature. Mother, earth, sun, light, children, family. It is all a chain of elements.

---

4 P.Capriolo, 2009 in Fotografia europea. Eternità. Il tempo dell'immagine. (Exhibition catalogue Reggio Emilia, 30th April-7th June 2009) edited by E. Grazioli, 2009, Mondadori Electa, Milan, p. 150-151.
5 P.Capriolo, op.cit., 2009.

A.M. In *Earth* (2006), on the other hand, the human presence occupies a more marginal role.

M.S. It is a work on different places, done at different times, while in *Ra* there is a rotation of the point of view, everything happening in the space of just a few metres, in the same place and at the same time. *Earth* is composed of photographs I took on the two islands of Fuertventura and Lanzarote. At certain points you feel as if you are looking at a lunar landscape. Going somewhere to work sometimes means coming across elements which then become the essence of the work. Not wanting to produce a descriptive work, not wanting to document places, I rely on what I find, in order to structure the story in a cinematographic key. I found stones which, in my opinion, might represent a kind of primordial chaos. It seemed to me the dimension resembled that of Land Art.

A.M. Does your work, in which man is a co-protagonist, have a social significance?

M.S. I think that a certain kind of beauty, a certain idea of nature and a certain type of image and colour actually channels a reflection on one's own existence and on one's own way of being in the world so as to return to the eternal questions: Who am I? What am I doing here? The element of contemporaneity is always emotional. This is not a contemporary landscape, à la American "New Topographics", of scorched earth. It is something which should pull you in instead. It seems to me that landscape, nature, both understood as the planet, is a highly social element. Nature should become an element of key interest to politics. Behind the sunset, behind the beauty of a dreamlike landscape, there are actually threats as well.

A.M. Walter Guadagnini wrote[6] that your work is founded on the ambiguous relationship between what is visible and what cannot be expressed, between what is seen and what isn't said.

M.S. In photography I am extremely interested in what is apparently not visible. By this I mean the visible which becomes the threshold to the invisible. The image and the subject become a medium for talking about something which is not the subject but something which lies behind the subject. Photography may contain something of a puzzle. It is like a kind of anagram. Things are photographed, they are reproduced and reality is recomposed apparently in the same place on film. Like a revelation. From this point of view I am also interested in the shape of the rebus, where you have an image that needs to be described in order for the rebus to be solved and so you draw on reality again. The rebus becomes a consequence which has no relevance to the image from the point of view of the meaning of the sentence. It is the passage between truth and reality, what Thomas Ruff called real reality, in the first degree. With photography I refer to a first degree reality. I photograph while I'm moving, travelling, walking. I am a street photographer. I don't produce images on the computer. Reality is always the starting point.

A.M. Standing in front of your works one might be tempted to speak of visionariness.

M.S. I think that it is rather a case of vision. In one of my texts which accompanied the exhibition *Idea di metropolis*[7] I wrote that the photographer is a passer-by who is always attentive to pre-existing points of view. I do not set vision in motion because I look and take photographs.

The vision is present and exists. The dream is the proof that it is not just by looking, by using your sight, that you can have visions. It is a constant deferment between other minds. Vision is a continuum between what I see and what looks at me. Mine is an attempt to reveal a vision which is already present. I like to search among the most banal things which are related to reality, seeking a significance which leads to something which goes beyond.

The landscape in *Earth* for instance is extremely lunar and you perceive a dreamlike dimension. I like a less structured, more desolate landscape. It is a horizon which we lack in Italy. Mine is an attempt to produce an effect which makes you reflect on time, which may seem something different to you than what I am showing.

A.M. Can we talk about the importance books have for you and their place in your project?

M.S. My books are not simply work catalogues but autonomous works instead. The idea of the book is fundamental. It is the work, the cinematographic piece, the film that allows you to convey the significance which might struggle to emerge in a single piece of work. I like working in episodes, which mean to be the essence of my various projects. I love books that are organised into stories. I like the fact that various collected stories manage to achieve a greater significance and to become a more structured story. It is like dispensing things slowly, letting the works settle, perhaps for years, so as to use them in a new key, just as we see happening in this exhibition. This operation offers up the real dimension of transformation. It provides us with the opportunity to reinterpret what has been done in the past and, in some sense, to refocus on it.

---

6 W.Guadagnini, Enigmi e incanti (catalogue of the exhibtion at Galleria Metronom in Modena, 4th April to 6th June 2009).

7 The exhibition took place in 2002 at Museo di Fotografia Contemporanea Villa Ghirlanda, Cinisello Balsamo (Milan).

Marco Signorini, 1962. Lives in Florence, Italy.

Le immagini di questo libro fanno parte di lavori diversi.
Les photos contenues dans ce livre font partie de deux projets différents.
The images in this book are from various works.

Earth, Fuerteventura and Lanzarote Islands, 2006.
Courtesy Galleria Manzoni, Bergamo, Italy.
www.galleriamanzoni.it

(Ra), Fanø Island, 2008.
Courtesy Galleria Metronom, Modena, Italy.
www.metronom.it

Altre immagini sono state realizzate fra il 2000 e il 2010 in Toscana.
D'autres images ont été réalisées en Toscane entre 2000 et 2010.
Other images were made between 2000 and 2010 in Tuscany.

Thanks to
Andrea Albertini
Michele Buda
Paola Capriolo
Walter Guadagnini
Danielle Igniti
Angela Madesani
Marcella Manni
Giacomo Manzoni
Emilio Mottolini
Eleonora Pasqui
Lorenzo Tugnoli
Roberta Valtorta

Questo libro è stato realizzato grazie al contributo del centro d'arte Nei Liicht, Dudelange, Lussemburgo.
Il centro d'arte Nei Liicht ospiterà una mostra personale di Marco Signorini a gennaio 2012.
Curatrice: Danielle Igniti, direttrice dei centri d'arte di Dudelange.

Livre réalisé avec le soutien du centre d'art Nei Liicht - Ville de Dudelange, Luxembourg.
Le centre d'art Nei Liicht va accueillir Marco Signorini pour une exposition personnelle en janvier 2012.
Commissariat: Danielle Igniti, directrice des centres d'art de la Ville.

This book is published with the support of center of art Nei Liicht, Dudelange, Luxembourg.
In January 2012 the center of art Nei Liicht will host a personal exhibition of photographs by Marco Signorini.
Curator: Danielle Igniti, director of the centers of art in Dudelange.

www.centredart-dudelange.lu

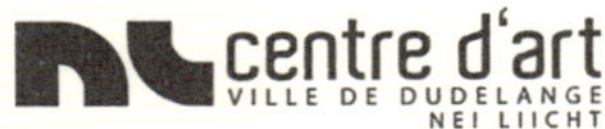

Galleria
Manzoni
Arti contemporanee e disegno industriale.

METRONOM

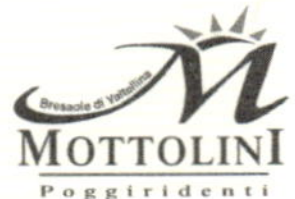

MOTTOLINI
Poggiridenti

MARCO SIGNORINI
EARTH HEART

© Damiani 2011
© Marco Signorini for the artworks
© Angela Madesani for the text

Editorial Coordination
Eleonora Pasqui

Translations
Joseph Denize
Judith Mundell

DAMIANI

Damiani editore
via Zanardi, 376
40131 Bologna, Italy
t. +39 051 63 56 811
f. +39 051 63 47 188
info@damianieditore.it
www.damianieditore.com

ISBN 978-88-6208-180-1

Printed in June 2011 by Grafiche Damiani, Bologna, Italy.

www.marcosignorini.it